The Search for a New Man

By

Thom Cantrall

COPYRIGHT PAGE
ISBN-13: 978-1490567846

SASQUATCH – THE SEARCH FOR A NEW MAN
COPYRIGHT 2013 THOM CANTRALL
ALL RIGHTS RESERVED
INTERNATIONAL COPYRIGHT PROTECTION IS RESERVED UNDER UNIVERSAL COPYRIGHT CONVENTION AND BILATERAL COPYRIGHT RELATIONS OF THE USA. ALL RIGHTS RESERVED, WHICH INCLUDES THE RIGHT TO REPRODUCE THIS BOOK OR ANY PORTIONS THEREOF IN ANY FORM WHATSOEVER EXCEPT AS PROVIDED BY RELEVANT COPYRIGHT LAWS.

WWW.CREATESPACE.COM
PRINTED IN USA
ISBN-13: 978-1490567846

COVER: ORIGINAL ART BY ALEXA EVANS

www.GhostsofRubyRidge.com/alexart

Dedication

There are three men in this endeavor who deserve the accolades of every person who loves sasquatch. These three men have done more that any others in the history of the search to not only bring the creatures to the public light but to ignite the imagination of generations of investigators.

Jerry Crew was a cat-skinner (bulldozer operator) building road in mountainous northern California during the summer of 1958. On August 25th of that year, Mr. Crew came to work on a Monday morning to find a line of tracks descending a very steep mountainside and then trailing around his tractor before heading down the road. In the soft soil of the newly turned road grade, the sixteen inch tracks sank deeply and were very visible and startling in their clarity. Jerry called the newspaper in Eureka, The Humboldt Times, where he spoke to Andrew Genzoli who wrote an article that was picked up by the wire services and appeared across the world. On this date, the world learned a new word. "Bigfoot" had been unknown prior to this moment. Even more importantly, a world renowned paleontologist, Dr. Ivan T. Sanderson, wrote articles that appeared in "Argosy" and "True" magazines which served to light the fire in a fifteen year old young man that lived not so far away. Without Jerry Crew, I might never have known the intense pleasures I have learned from dealing with these beings.

A short decade later, two cowboys from the Yakima, Washington area rode the very roads that Jerry Crew built to history. Roger Patterson and Bob Gimlin were in northern California in an attempt to find photographic proof of the existence of this enigmatic and elusive being. The film shot by Roger never provided the definitive proof he had hoped for, but that definition is not possible in the climate of disrespect that pervades the world today.

To these three men, I dedicate this volume and the stories contained herein. I thank you, gentlemen.

Prologue

This part is most special to me... Bob Gimlin is my friend. I have spent a lot of time with him talking about this incident as well as just talking about life. Bob is one of the most interesting, most forthright and most honest men I have ever known... and I have known church leaders, giants of industry and leaders of men. Among them all, Bob stands out as a real man. This is his story as he told it.

An Interview with Bob Gimlin in 2011

By

Thom Cantrall

The reason for the expedition was simply that Roger Patterson had caught the bug... he wanted to do a documentary on sasquatch. He and Bob were buddies. They had rodeo'd together...

Roger Patterson with casts

traveled together a lot. So he recruited Bob to go with him.

They had been scouting the Bluff Creek, California area for several days. They rode their horses wherever they could get them. The country is very rugged there and much of it is inaccessible except on foot. It is not "horse friendly" at all in most cases.

(This part is directly from Bob Gimlin himself) On the twentieth of October they had been riding on the ridge and came on the tracks of seven creatures... they followed them and when three of them went down off the ridge to the creek bed... they followed... They were riding up the right side of the creek when Patty appeared... Roger's horse went ballistic... just blew up on him, and he bailed with the camera. He ran across the creek trying to film...fell down while trying to climb the bank on the left side... He finally reached a spot where he could stand and started filming. Bob told me that the famous frame 352, where Patty turns to look back at him is when he crossed the creek and stepped down off his horse with his rifle drawn, but not shouldered. Roger had asked Bob to "cover him" as they knew the rest of the clan where very close. She was NOT there alone.

Bob wanted to follow her on horseback, but Roger said, "Don't leave me unarmed." His horse was not available and his rifle was in the saddle scabbard. Therefore, Bob stayed with Roger and they examined the trackway and measured it out. Bob told me something that I saw with my Dickey River friend as told in Chapter

Three of this volume. "Her tracks across that flat were slow and measured. She was just retreating..." He told me she had a juvenile in the brush, just out of range of the camera... in fact there is one short spot where you can see it if you know JUST where it is. "But," he said, "as soon as she was out of sight, her stride lengthened considerably." As she departed, she doubled back, recrossed the creek and ascended the steep bank on the right of the stream. That is the same thing Dickey River did to me. He just sauntered out of my view and then his trackway showed a HUGE increase in stride... he was booking from there!

The reason the creek was so open then was that just a very few years before, in 1964, there was a MAJOR flood across all of N. California... it was classified as a "100 year Flood"...

Now, let us hear the story of this day from the man who was there on the twentieth of October, 1967, Mr. Bob Gimlin...

Presentation from Bob Gimlin Concerning the Origin of the Patterson-Gimlin Film

By

Thom Cantrall

From the 2010 Ohio Conference

"I thank all of you for being here… I thank Don for inviting me here."

"I had a little accident. If you guys don't mind, I want to tell you a little bit about myself. I won't whine a whole lot 'cause they say cowboys aren't suppose to whine but let me tell you when I first got this it hurt and it still hurts."

"So, I wasn't sure I could make it. When it first happened, the first couple of weeks I was in so much pain it bothered me a bunch. And the doctor said, 'Don't you dare do it.' I got to thinking about it. Of course, some of the folks I've known before were coming here and I have some people that are practically my relatives came up so I figured if I don't show up, I don't know how I'm going to answer to them."

"I haven't had a chance to meet each and every person but I sure would like to. I know it's gonna be impossible, but if you folks will allow me, I'll tell you a little bit about my life. I come from Yakima Valley, Washington which is way on over there. All the bad stories you hear about Washington aren't really true… maybe some of them."

"But anyway, I basically ride horses for an occupation and you can see that I probably should change occupations. And, I intend to."

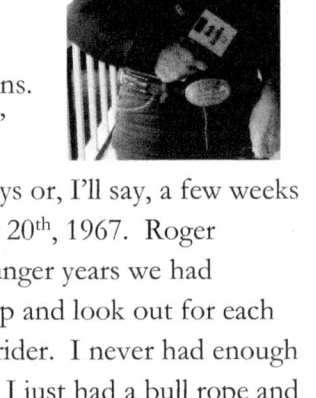

"So, what happened to me a few days or, I'll say, a few weeks before this big event happened on October 20th, 1967. Roger Patterson and I were friends and in our younger years we had rodeoed together. Cowboys kinda buddy up and look out for each other… especially if you are a rough stock rider. I never had enough money to buy a good horse and a saddle so I just had a bull rope and a bareback rigging, if any of you folks can understand what I'm talking about. So, consequently, that's what we did."

"Then there was times when my family started growing a little bit so I figured I'd better quit riding rough stock and the rodeo circuit. I hadn't seen Roger in quite a few years. One day he pulled into my driveway and he had a great big old plaster cast with him. He said, 'Bob, do you know what this is? And I said, 'No, I didn't.'"

"But, I'll go back just a little ways… I got married in 1964. My wife and I went to Harrison Hot Springs, BC (on our honeymoon). We got up there and as we were pulling out of Harrison Hot Springs, you look over there and the was a big plaque with a sasquatch on it and it said 'Sasquatch'. My wife said, 'Sasquatch'? What's a sasquatch?'"

"I said, 'I don't know, just a big ugly Indian, I guess.' That's when I first heard the word sasquatch or knew anything about it."

"When Roger came flying in my driveway that time in his little Volkswagen Bus and showed me that plaster cast, he said 'Hey, that's a bigfoot track.' And I said, 'yeah, it is pretty big.'"

"Then he brought a book down by Dr. Ivan Sanderson. I don't know if any of you folks have ever been able to get ahold of that book. It's a real good book in my opinion. Anyway, to go on with my story, I read the book a little bit once in awhile when I had some spare time, which wasn't often. Roger kept coming by. He lived up in the woods quite a little ways from me, about twenty-five miles. I was riding a lot of colts… young horses, that is. I'd drive up past his place and I'd stop and visit with him. He always had a story and he had these little cassette tapes and he'd play them… about testimonials of people that had… well they weren't witnesses of bigfoot, but they had some kind of episode with them… some kind of encounter that they didn't understand or whatever, you know? So consequently, I got to the point where I'd sit around the campfires at night and get to thinking, well maybe there is something out here that's making all these people make these statements and Roger with these plaster casts that he'd cast in different areas."

"He had made some trips to northern California also. He was in it pretty heavy, but everything went along. I really didn't have much interest because I had too much… too many other things to do to make a living. I had four children at that time so consequently I was working pretty hard to keep up with the everyday bills and so forth. As time kept going on by, I'd think about this and I'd go past Roger's and I'd stop and pick him up and his horse. We'd sit around the campfires at night and he'd tell me all about these things. I'd listen, 'cause that's all there is… not a lot to do when you're around a campfire at night. Roger hollered at me one Friday I think it was… Labor Day Weekend… and said, 'Hey, let's go over to Mount Saint Helens area and ride.'"

"I'd never been there. Mount Saint Helens is beautiful. And he talked about Ape Canyon and the Beck deal and so forth. I

Mount Saint Helens prior to 1980 eruption

was listening to all that and thought it was a pretty good deal. So, we went over there to the Mount Saint Helens area and found out you couldn't ride your horses very much in that logged off country because it was just slash and stuff down everywhere."

"It started raining really hard over there on Sunday so we came back on Monday and I went back to work. About a week later Roger came flying into my place again and says, 'Bob, we gotta go to northern California.'"

"'What for,' I asked"

"'They found tracks down there around a piece of equipment or a tank that they pulled in there. There were three different sizes of footprints. They don't really know if it was a male and female and a young one, but there was three different sizes of footprints.'"

"I said, 'Well, Roger, I can't just go right now.' This was like just right after Labor Day. I said I got things I've got to get lined up here and my wife wasn't too interested in taking care of a bunch of horses and cows. She had a full time job also. I finally talked her into doing part of it and got ahold of some young people to help while I was gone. I wasn't going to be gone very long. I just figured another few days."

"So we get down there to northern California. I'd never been there before so I didn't really know where I was going. We get down there to that area and Roger, of course, told me how to get in there. He'd been in there prior to that. He'd talked to some of the people down there, Al Hodgen and a few others... So, we went in there and camped."

Bob Gimlin with pack horse

"By the time we got there these footprints around this piece of equipment that was new faced dirt were just blobs of mud... I mean blobs where it had rained... it had rained hard down the whole west coast, I suppose. There wasn't even, as far as I was concerned, an identifiable footprint there... not one that you could put a cast or put plaster in and get a good cast out of it. I wasn't still impressed and up to that time I had never seen a footprint that you could get a cast out of so, you know, I still wasn't a confirmed believer in anything."

"We rode those mountains every day with the horses... around and around and, you know, just covered a lot of miles. I had a truck, a dual wheeled Chevrolet truck that was not a four wheel drive, so it wasn't easy to get it around. They were pushing logging roads in up through there to start logging way up above where we were camped. So, I'd take that truck and we'd drive the roads at night to see if we could see any footprints... tracks cross there for footprints. We seen everything but a bigfoot track... deer, cougar, bobcat and bear."

"This went on for quite a few days and I was getting ready to come back to Yakima. Roger said, 'Let's go up to that area that really looked like great habitat a couple more times and then maybe you can just leave me down here and come back and get me in a month.'"

"I said, 'No, if I leave you down here, you're gonna be here, I'm not coming back after you!' It's quite a little ways from where I lived down to where this film footage was taken."

"We left out that morning fairly early... Well, go back a little bit, I left out early. I always got up early... Roger didn't. He usually slept in a little more than me. I'm kind of a farm boy so I got up about daybreak every day and saddled my horse and I'd ride out. Well, my horse loosened up a shoe so I rode back in to tighten... to get the equipment to tighten the shoe back up and get it going. Roger came in just a little bit later and he said, 'By golly, Bob being as your gonna go in a few days here, can we go out and stay all night up in the mountains in some of that area where it really looked like great habitat?'"

Patterson-Gimlin Frame 352
Used with permission

"I said, 'Well, yeah, I suppose.' So we started out that afternoon, well, it was early afternoon... it suppose it was twelve... twelve-thirty... something like that. (We) took a little horse with us with our sleeping bags and our equipment all on it... rode up this creek bed about three and one half miles from where we were camped... Came around a bend by the crick where there was a ... I don't know how many folks here understand when I say downfall tree. There was a tree downfall with the roots system up. It was probably eight to ten, twelve feet high and that had caused the crick to reroute itself and go around there. I'm not real good on diagrams like Jeff and some of these other guys are. ... I'm not comparing myself to Jeff... I'd never even try that, let me tell you! Anyway, when we came around that bend in the creek, the creature stood across the crick about as far as far as that gentleman is right there with the paper in his hand (indicating a very close encounter). It was standing when I saw it. It may have been stooped down when Roger first saw it, but I was just a horse length or so behind Roger, leading the packhorse. When I saw the creature it was standing up looking across the crick to where we were. But, it immediately turned and started walking away... just taking natural steps. I mean natural looking steps to us which measure out forty-two to forty-eight inches from heel to toe... which is a pretty good step. There are some pretty tall men here and I'd almost dare them to make a step comfortably that long and keep on walking."

"That's how this film footage kinda got started. Roger's little horse kinda threw a little... I don't know what kind of dance you'd call it, but he wasn't... Roger was trying to bail off and get his camera at the same time... which he did. Roger was a very agile man... of course, he was a rodeo cowboy and he was a great athlete. So he bailed off that little horse, with his camera and ran across the crick trying to get that camera focused on the creature. There was a little incline of sand alongside the crick on the other side. Roger kinda stumbled and fell when he got on the other side of that. He went down on his elbows. If you've seen that film footage from beginning to end, you'll see all that shaky part at the beginning. There is hardly... you can't hardly identify anything. Then Roger realized the creature was moving on and all this time I was setting right where I was. I stopped... the packhorse pulled loose or I let him go, I don't know. Everything happened so rapidly. I really don't remember if he pulled loose of if I just threw the rope back at him 'cause I was having a little hard time holding my horse. It was an older horse.. it was an old roping horse and he was a little bit easier to handle than Roger's little horse that had a lot of spunk."

"There's when it all started going kinda wild. Roger wanted to relocate because she was movin'... I say she but I didn't know if it was a female or a male... it didn't make any difference, it was going,

PGF site X marks frame 352 site

you know? Roger had to relocate to get a better... a better view, so he asked me if I would cover him. He told me later that he never... he was kinda concerned about two more being right close to this one. It was a wooded area on the far side... a really wooded area and you couldn't see up in there so that's what his concern was. Well, I didn't know that but he said 'Bob, will you cover me?' Well, I know I had a rifle in my scabbard on my saddle but I knew I couldn't do it sitting on a horse if anything happened that I needed to do because I was an avid hunter at that time and shot a lot of big game. I knew what I had to do, so I rode across the crick, stepped off the horse with my rifle in my hand... and that's where you see that famous turn when she turns and looks back at me... when I stepped down off that horse."

"Maybe the misconception was that I intended to shoot this creature... I did not intend to shoot the creature. I never raised the rifle to my shoulder... ever. I was carrying a .30'06... a Remington

.30'06 with 180 grain bullets in it. I felt if I had to I probably could stop this creature if it came back at us. As long as she kept walking away, I had no reason to even bring the rifle up. Like I say a few seconds ago, everything was happening so rapidly you just didn't have to do a lot. She was walking away all this time. Then, Roger hollered at me, he says 'Damn, Bob, I ran out of film...'"

"I got back up on the horse, I said, 'I'm going to follow...' and he said, 'no... no, no, no, no, no... Bob, don't do that, the other two might be here...' So he was assuming this creature had... we were only about four miles from where those footprints were... Roger was assuming there would be two more there and he didn't want to be left there with just a camera in his hand... and no film in it. I mean now it seems kind of amusing to us but I don't think it was to Roger... when I seen the look on his face he was glad I turned around and came back."

"By then, he got under a poncho. I didn't know much about cameras or anything... got underneath an old poncho that I had on the back of my saddle and got more film in his camera. Then we proceeded to catch up his horse which ran

Camera Model Used by Roger

down the crick a ways and the little pack horse. And then we tried to follow where this creature went. We got up there a quite a little ways and we only saw one half of a wet print on a rock going across the crick and up the side of the mountain. It was steep and I kinda wanted to go after it, but I didn't know really why, I just wanted to see it again."

"Roger said, 'No, we gotta get back, Bob.' 'Cause October twentieth your days are pretty short down there in the mountains."

"When that sun sets over them hills it don't take long 'fore it gets dark. By then, it was in the afternoon and we had to go back down to camp to get the material to make the casts. We did some things there. Roger had me get up on a stump about, oh, three and a half or four feet high maybe… that was alongside of where she went and made her tracks. I had on a cowboy boot with a riding heel which is a pretty sharp heel now if you understand what that means. At that time I was a little heavier than I am now… I weighed about one hundred and seventy five pounds. I'd jump off beside this footprint… tracks… with that heel with one foot hit first and see how deep I could go into the soil that she walked through. It wouldn't go near as deep as her tracks were. Then I rode this horse alongside. He was a sixteen hand quarter horse and he weighed about twelve hundred pounds. With my weight on him and the saddle I rode him right as close without disturbing the tracks as I could. Roger took pictures of that. The horse never made tracks as deep as the creature did. So that indicated right there that she was fairly heavy… that she was a heavy, heavy muscled creature which you can see in all this. People say, 'What'd she weigh, Bob?'"

"I don't know I thought three or four hundred pounds was really big, you know? I said, 'probably three or four hundred pounds and, and a half foot tall.' Well, come to find out, I was way off on everything, you know? But, like I say, when I first saw her I was up on a horse sixteen hands high which I was about nine feet up. Things will look a little different from that height."

"I'm going to finish this up by just… and you folks see what the result was from that film footage… that short film footage that we were lucky enough to get. At that point in time being as Roger fell down, we had no idea that we had any good film footage at all. Naturally we got the cast made and the pictures made of what we hadto do there for what evidence we could get. Then we went in to mail that to Yakima or wherever he mailed it to. There has been a lot of controversy on where that film was processed and where it was mailed to. I never paid that much attention to it because I was very tired from being down there two weeks, riding horses every day long hours and driving the truck at night."

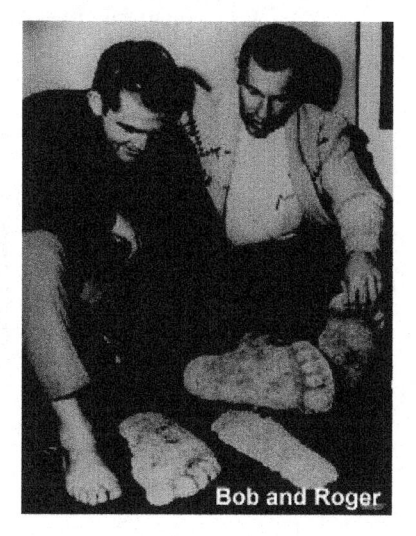

Bob and Roger

"Roger slept in a whole lot better than I did cause I just didn't sleep that much down there."

"What I'm going to do now is finish up because I don't want to be taking somebody's time that means more to you. Folks I sure appreciate this. I'm open for questions but I can't hear because I'm such a young feller. But, if you want to, Don (Keating) can be my ears and I'll answer anything I'm capable of answering for you if I haven't forgot over the years."

Q: How far were you able to track it after it disappeared?

A: "We was able to track it through the entire area where it walked through and then we measured scuffs in the gravel along the crick where it went up the crick. They weren't really accurate. We could only measure how the gravels were scuffed and turned over. Those were Sixty-Two to Sixty Eight inches apart in stride. So it appeared… speculation was that once she got out of our sight, she ran."

Q: (How far did were you able to track her?)

A: "Well, probably two hundred yards. From wherever Roger first started filming her probably a total of about two hundred yards because that was where the end of the good soil quit and the gravel started. You could track but there were no identifying marks."

Q: Is it true you and Bob Heronimous are still friends…

A: "Well, Henry, a friendship is something pretty valuable. I still speak to Bob Heronimous when I see him… I don't consider him a friend."

"I wanted to say, there was only Roger Patterson and I down there at that time… there was no one else down there."

Q: I want to thank you for not shooting this animal. I know they're out there and I want to thank you for not shooting it. We don't need to be shooting these animals.

A: "I totally agree with you."

Q: Did she appear to be afraid of you at all?

A: "No, not that I could detect. You know, when she walked away, she never did run… and I keep saying she… at the time… IT never did run until it got out of our sight and then we assume that she ran because the distance between the scuffs in the gravel. We knew what were hers… she went that way and they were fresh… the only gravels turned over with moisture on them."

Q: Where is Roger Patterson now?

A: "I hope he's in Heaven."

Q: Did Roger think it was a female at the time he saw it?

A: "We didn't even discuss it because this happened so fast. I never knew it was a female until everybody started looking at the film and said 'Well, she has mammary glands so it must be a female.' So, that's when... when they first started talking about Patty, I had no idea... I thought they were talking about Roger Patterson's wife... That's Roger Patterson's wife's name."

Q: Mr. Gimlin, we've all heard the stories of people that have shown up saying "I was the man in the suit..." I even saw once on tv it was a woman in the suit. The only thing I can say is, "Where is the suit?" No one has ever shown us the suit, do you know what I mean?

A: "Yes, I know what you mean."

Q: (Something about why he didn't say more... that she would be shouting from the housetops to let everyone know)

A: "Have you ever heard of the 'Day of Hell'? That's what it was for awhile. Well, there was so much going on back then. It wasn't people like you folks here that I appreciate you being here... In my area, you know, a bunch of people... farmers and such and they weren't concerned about it and probably didn't believe it. There's an awful lot of things that went on right after that film footage. I went home... went back to work and Roger and Al traveled with it and did things with it. I, um, I just took an awful lot of static over it. Even my own family kinda… you know… They knew I was a little different from the rest of them…"

Q: Bob, were you under the assumption that this was Patty's first encounter with humans because of the way she acted? She acted real nonchalant and real casual as she left the scene.

A: "You know, I wish I could answer that but I couldn't do it honestly because I have no idea, you know, how these creatures react when they see humans and horses. And in that area, there had been quite a few of these sighted in that area before. There had been footprints got there a lot. In fact there was the song that Tom (Yamarones) has here about Jerry Crew(s), excuse me. There's been a lot of footprints cast down there… different sizes like some of these that Dr. Jeff (Meldrum) has here. I don't know if any of these are those, but at that one time Jerry Crews had about fifty different footprints and shapes and he lived down in that area."

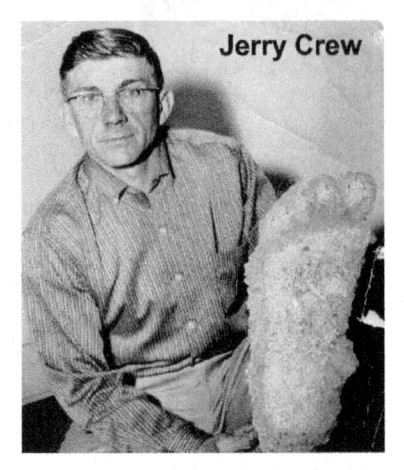

Q: Do you know where the original film is today?

A: "I do not. There's been claims… somebody said it's … it's there… and, you know, I had no idea because Roger had total possession of that. Between him and his brother in law, Al Depp."

Q: In your professional opinion as a horseman, do you think the horses would have reacted as strongly if you'd come across, let's say, a grizzly bear?

A: Well, I hadn't ridden this horse I was riding very much, but I have… We don't have grizzly bear where I'm at. We have brown (black) bear. So, I've ridden a lot in the mountains and encountered brown bear and it kinda depends on what the bear reacts and what horse you're riding. You know, whether you're riding one that's been around a long time or one that hasn't been around a long time. I have ridden horses up in there and brown bear go 'woof, woof, woof' and starts running down the trail and the horse kinda throws a fit and has a little trouble and you gotta be...

you gotta stay with them a little bit."

Q: Do you think we should be afraid of these creatures?

A: "No, I do not, honey. I think if we leave them alone... they are curious creatures... This is my own personal opinion and I strongly believe that these creatures will not harm human beings if they are left alone. But, I think if we go out there and shoot at one and wound it, it's like any other animal that's been wounded... you might have a problem on your hands."

Q: Can you describe the circumstances when you first saw the Patterson film... who you were with, etc.

A: "Yes, I was so tired... I drove... I couldn't even go into what happened right after the film 'cause it'd take another hour. When I got home... I drove all the way that night... drove all that afternoon. It took us all day to get out of there... it was raining... mudslides and so forth and so I drove like almost twenty four hours straight or so... it was actually more than twenty four hours and I was so tired I went to bed and I went to sleep. So, I didn't see that film footage for two days and when I did see it, my first opinion was, 'That's not very good, I seen it better than this.' Which is a natural assumption, you know because I did see it better than that. You see what I'm saying? I thought, 'Well, I don't know what everybody's getting excited about and all that, that ain't very good.'"

Q: (Have you seen another creature like that one?)

A: "No, I have not."

Q: Would you want to?

A: Oh, if I get a chance. YES… I… I… I… would really like to. I'd like to be able to walk up to one and look it in the face and smile at it and say, 'Hey, he he.' … if it was a male."

Q: After the filming, why didn't Roger ever return to the film sight for more evidence?

A: "Well, I can't speak for Roger, but I can speak for Bob… We…Roger and Al, they just could not leave well enough alone. And, this is just me talking now… because I wanted to stay there in Yakima and really just kinda… just see what happens really. Well, Roger and Al, right away quick, they wanted to travel with this film, which we did, and it wasn't really anything I cared about. First we went to California and talked to David Wolper, Wolper Productions. Talked about getting back up in there and doing some filming… doing some film for the agent. I got David Wolper mad right at the beginning because he said, 'Well, we'll send an actor up there and say he's Bob and send him up through there. I said "Hey, I want to tell you something Mr. Wolper,' and people don't talk to a movie producer like that but I'm just an old farm boy so I said, 'You don't have a man that can carry a camera and follow me up through there that will film me.' And he said, well, I think I do. And I said, 'Well, get him out here but you don't have, I can tell you…' because I was in great physical shape at that time and I could move."

Q: Could you smell anything different?

A: "Now, wait a minute, what do you mean, 'smell different'? Yes, I did. There was a smell. Roger and I never agreed on the smell… not totally agreed on it. I thought this creature smelled kinda musky… skunky like. Not quite as sharp as you smell a skunk as most of us have at one time. And Roger… he says it smelled like an old wet cow dog that'd been rolling in the cow manure. Oh man… don't wanna go there."

Q: Is there anything you saw on the creature that you cannot see on the film?

A: "Well, you know, that's kind of a difficult question but I'll do my best. The only thing that I could see that I don't see on the film, excuse me… as the film is the rippling of the muscle underneath that hair as it walked away. And, the sun was over there in the West and setting on us, so you can see at different see at different times, or I could see at different times that tremendous amount of bulk muscle down the back… in the thighs and the shoulders. And, everytime I've seen that on the film, I've never really appreciated that bulk of muscle I saw that October twentieth on that afternoon."

Q: How tall do you guesstimate it was?

A: "Now, that's a loaded question. At that time… that particular time of shortly after, I was asked that question. Well, when I first… when I first saw that creature, I was sitting up on a horse

23

sixteen hands tall and with my eyesight, that's an estimate of nine feet. So, I would have thought at that time that it was probably six and a half feet tall, but then, it was further away when I stepped down off the horse and things were happening so fast that I never really thought about the height. I just knew it was a big, heavy muscled creature that was big. So, I never realized when they asked me that... what I was really saying... I said 'Oh, probably over six foot tall. And, so, when Bill Munns done his study... when he did his thing on it, it figured out to be seven foot three and three quarters."

Q: How would say how broad the shoulders were on it?

A: "Biggest football player I've ever seen."

Q: Hey, Bob, prior to the sighting, did the horse act nervous or agitated or anything? Or anything like that?

A: "Not that I can remember that much, because, you know, I'm used to riding horses that act every way which you can tell. And so, a horse has got to get pretty stirred up before it bothers me very much and Roger was riding along in front of me and I never noticed his horse acting up although it was a real lively type horse. You know it was one of those sunny shiny October days and Roger was filming the red and colored leaves from October colors and the trees there. You know it was just a pleasant day and of course, things were kinda going along just perfect. I thought life was just perfect that day until this happened."

"You know folks, it's so great to have you here after all the years I took flack. And so it's been so good to see you here and I'd like to be able to talk to each and every one of you… I will as much as I can. And I'll be here the rest of today and tomorrow. Thank you very much."

Table of Contents

Dedication		Page 3
Prologue	Bob Gimlin's Story	Page 4
Chapter 1	Winter Wonderland	Page 28
Chapter 2	My Ozette Lake Family	Page 38
Chapter 3	Who is Sasquatch	Page 49
Chapter 4	Del Norte Encounter	Page 56
Chapter 5	Butterfield Canyon, UT	Page 68
Chapter 6	Revelations of Novelty	Page 73
Chapter 7	Todd Neiss Encounter	Page 83
Chapter 8	One Summer Day	Page 97
Chapter 9	Kathi Blount Report	Page 102
Chapter 10	Mindspeak	Page 112
Chapter 11	Vision – How We See	Page 121
Chapter 12	Intermembral Index	Page 128
Chapter 13	The Humanity of Sasquatch	Page 133
Chapter 14	Coppeii Creek Homesteaders	Page 145
Chapter 15	Voices on the Lake	Page 151
Chapter 16	H.E.R.O. Report	Page 157
Chapter 17	Parts of the Whole	Page 169
Epilogue	The Time Is Here	Page 176

Sasquatch
The Search for a New Man

The Story of the Emergence of a Hitherto Unknown Primate Human Species

Winter Wonderland

In 1968 the U.S. Navy had given me my choice of duty stations and I had chosen the Polaris Missile Facility at the Naval Ammunition Depot near Bremerton, WA. The duty here was superb as I knew it would be. Our job here was to test and tear down Polaris Missiles returning from duty in the fleet and build up and test those missiles going out to the fleet. Our staffing and work load was such that I had plenty of time off and I loved it!

I lived on a small island on the Olympic Peninsula that was an adjunct to the Depot and had full run of it for my own private recreation ground. The fishing and clamming was second to none. The deer made gardening a frustrating and useless gesture and the proximity of the Olympic Mountains, the Coastal Rivers and the

Pacific Ocean made this a virtual paradise. My passions at the time were elk and sasquatch. I was in a place where I could fully indulge my love for learning more of these huge, enigmatic beings called sasquatch.

I had met and befriended several people that became very important to me over the next decade or more. Most of them refused to accept the possibility of even the existence of this creature, so I learned early on to not even mention him in their presence. My wife and in-laws were in this group. My hunting partners were another part of that group. Even the fellow in this outing refused, after the event, to admit that the experience even happened. He maintained from that day to the day he died some few later that he had heard a bear and had seen no prints in the snow.

That this essay is factual goes without saying. Yes, it is written with a bit of a "tongue in cheek" attitude, but the events and circumstances outlined here are totally factual and occurred in January of 1969.

Chapter 1

Winter's Wonderland

By

Thom Cantrall

First off, let me say that I am pretty much a "live and let live" kind of guy. This is especially true of God's small creatures. If I am out hiking and happen upon a rattlesnake, I'll simply back up, tip my hat to him and wish him happy hunting. The fact that for the next hour or so I jump about a foot off the ground if even a small branch should happen to snap against my leg in no way compromises my calm demeanor in these matters. It's simply that in some things, what the brain knows logically is not necessarily retransmitted to the reactive nervous system, let

Paul Bunyan and Babe

alone to the feet and legs that cause these leaps of unfaith.

Probably the worst case of this "Induced Reaction Syndrome" as I like to call it occurred in a good friend of mine who is now passed on. This latter fact making it much safer to relate the tale as it occurred, without fear of reprisals or at least, a swift kick to the posterior.

Frank and I had decided to take advantage of the late elk season on the very north end of Washington's Olympic Peninsula. The weather was ideal for this January outing... snow, snow and more snow. It snowed all day the day of our evening departure from our Port Townsend area homes. By the time our gear was loaded, the trailer attached and we were on our way, there was more than a foot of fresh, white tracking snow on the ground with more on the way. Frank, being from Missouri and unaccustomed to the rigors of tracking elk while ploughing through hip-deep snow, was positively jubilant at the prospect.

Throughout our four hour drive to the Hoko River country, normally about a two hour drive, sans snow, he regaled me with tales of his misspent younger years in the "Show Me" State. If I were believe one-tenth of the antics he related to be "gospel true", his companions had to have had intellects such that, by comparison, an earthworm would be considered an over-achiever and a cucumber could graduate high school with honors. Some of these tales made Paul Bunyan and his Blue Ox, Babe, seem absolutely plausible by comparison. All in all, though, it was a delightful drive through falling snow with Nirvana waiting for us at the end of our trek.

It was fully dark by the time we arrived at our planned destination, the end of a logging road that led to two clearcuts, one freshly logged and the other about three years old, affording perfect feed for many head of deer and elk. This little road wound its way up the mountain a mile or so to the older unit, then on through standing timber to the freshly logged unit at the end of the road. My original plan had called for us to drive in on this road past the first unit to a wide spot by a small stream where we would set up our trailer. Clallam County Road Dept., being reluctant to expend time and effort clearing private logging roads, forced us to alter these plans and camp just off the paved road, blocking all access to the snow-choked logging road. After setting up camp which, for us, consisted of unhitching the trailer, making it somewhat level fore and aft, turning on the gas and lighting the pilot lights on the stoves and refrigerator, the work of about twelve minutes flat, we retired for the night.

I should note at this point that the reason for choosing this particular area was that a good friend of mine had logged that back unit and had seen elk nearly every day feeding in the front unit just at daylight. Our plan was to be in that unit just above the Hoko Road well before first light and see if we couldn't ambush one of the big Roosevelt Elk that live there. To that end, we were up well before daylight, had a bite to eat and were headed up the road, climbing the mountain in the diminishing snowfall. Just about the time we were leaving camp, our bowstrings safely tucked into an inside pocket where they would stay warm, dry and serviceable, the falling snow changed to rain. It was not a heavy downpour like that ultra-wet country is capable of generating, but a soft, steady drizzle. It soaked everything... Snow... Trees.... Hunters....

On and on we slogged through the white expanse, climbing the steep grade that led to our goal. Although there was no moon out, the glow off the now melting snow afforded us ample light to see into the dark night without the use of flashlights. Careful we were to not get off the edge of the roadway for a fall into the adjacent canyon in the gloom of night could prove fatal. We were reluctant to show any light lest it be visible ahead of us and into the clearcut.

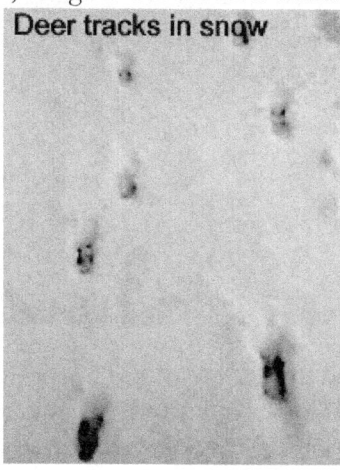

Deer tracks in snow

About thirty minutes before dawn, we reached the edge of the clearcut and decided to wait there where we could see the entire unit and await daylight. There were no tracks on the road, but we had expected none as we had the end of that road blockaded by our camp. The light, steady drizzle was doing its very best to work up into a full-blown downpour and the temperature had risen to well above freezing. That alone would make any tracks encountered to be very recent, indeed. In fact, I don't believe a track in the open would be at all crisp and delineated after as little as a half-hour in these conditions.

Slowly, the skies began to grow brighter with the promise of a new day, a new creation, even. Stumps began to emerge from the black of night to belie Frank's profound belief that they were a herd of elk feeding in the pre-dawn air. Just as one particularly majestic six-point bull disintegrated into its component parts consisting of a very nice Western Hemlock stump, a short length of cull log left as useless by the loggers and now sticking out of the snow at just the right angle to make a beautiful elk body topped with an advantageously placed branch pointing skyward in just the right place.

While watching the disintegration of the nice bull, a deer came off the bank behind us and walked slowly and quietly between Frank and me at a range of less than thirty feet. It was as if this deer knew that his season was closed and, hence, he was safe. More likely, the heavy air and falling rain did not allow our scent to travel far and I doubt he ever knew either of us was there.

He was not a large buck, merely a forked-horn, probably a two-year-old that weighed not more than a hundred pounds soaking wet, which he certainly was, along with everything else in these environs. These Columbian Blacktail Deer are not large deer under the best of circumstances and this particular animal was a youngster to boot. Slowly and cautiously he ventured step by precarious step past us and into the snow covered brush that was the clearcut. He was being extra cautious in his trek. I assumed then it was because of the poor conditions. Many times I have seen animals in similar diminished conditions behave in a like manner. When scenting and hearing conditions deteriorate, they become ultra-wary and extremely reluctant to trust their usually keen senses. Often they will lay up tight and not move until conditions improve for them.

When this little buck had placed enough distance between us that he was beginning to blend in with the stumps, he suddenly came to a stop, his head up, his ears erect as he stared into the darkness before him. What he was seeing, I had no clue. But, seeing something he certainly was. As he peered intently before him, he slowly raised one front foot then quickly stomped the snowy ground before him while emitting a quick snort through his nose. This is a behavior I have observed many, many times when a deer has spotted something out of place before him, but cannot decipher what he is seeing. Personally, I believe it to be one of two things… either it is an attempt to get whatever it is seeing to move, the easier to identify it… or, it is a warning to other deer in the area that something is amiss. This would be similar to the stotting of the Mule Deer, the stiff-legged bounce that can be heard for a considerable distance, putting every deer around to flight. Perhaps it is a combination of the two, but whatever its purpose, it was effective in this case as there came a "woof" out of the night sounding like, but not precisely the same as the huffing of a bear as he feeds his way among the rotted logs and such. I heard it quite clearly and I have heard the wuffing of many bears and the point must be made that, while it was reminiscent of that, it was not that! It was enough different that I immediately shot straight up off my stump/seat. The deer, too, was alarmed as he wheeled quickly and sped back towards us, passing so close that Frank had to dive behind a stump into the snow to avoid being hit by the escaping deer.

Frank hurried over to my position as fast as he could negotiate the snow. "What in Holy Hell was that?" he yelled loudly with his eyes approximately the size of dinner plates.

"Shush," I admonished him. "I don't know what it was, but I do know that deer didn't like it, so I think we should stay right where we are. It will be light enough to see within the next thirty to forty minutes to possibly an hour, depending on the density of the cloud cover. We can see what's up then.

Reluctantly, Frank retraced his steps to his stump and resumed his vigil. It was a vigil, he later related

Distinctive Double Tracks

to me, "that took so damn long that I was sure I'd have a beard down to my belt line before it ever got light!"

When light covered the land sufficiently, I whistled to him to come to my position. When he got there approximately one and a half seconds later, I was moved to ask if he had been shot there by a large rubber band. Frank was shaken, I could see that. It was so much so that I asked him if he wanted to go back to the trailer. His head bobbing up and down so hard that I thought his wool stocking cap was going to shoot right off his head told me that he'd done about all the hunting he was up for on this morning. I told him to just follow the road right back down the mountain and he'd run right into camp.

"You're not coming?" he asked plaintively.

"Oh, heavens no," I answered. "I came out her to arrow an elk and the conditions are almost perfect for hunting, so I'll be danged if I'll quit now. Besides, I want to find the tracks and see what it was that made that sound."

"Damn it, Thom," he pleaded, "don't do that! It's just too spooky. Let's go back to camp now."

"Go ahead, Frank," I suggested, gesturing down the trail, "but I'm going on. There are elk here. I've caught their odor a few times this morning and I want one."

With that, I simply turned and started walking up the road, intending to cross the cutting unit and then glass it from the far side. Also, if my deductions were correct, whatever had made that noise should have crossed the road either coming or going. I was a bit amused to hear a thoroughly exasperated Frank immediately behind me... so close that if I'd have reached into my hip pocket for a handkerchief, I'd have shaken hands with Frank!

Sasquatch Print

We had moved less than a quarter-mile further into the snowy expanse when I spotted something on top of the snow just ahead. As I made my way to it, I wondered what it could be. What I discovered amazed even me. What had caught my eye was fresh mud on top of the snow. And, what had been the source of the fresh mud were fresh tracks in the snow... Humanoid tracks... approximately seventeen inches long and a third to a half as wide. The distinct impression of five very human-like toes so clearly defined told me that these tracks were no more than an hour old, probably less.

Investigation told the story. This humanoid creature was carrying/dragging something with him. The hairs I found indicated it was, most likely, a deer. At the base of the fill over the culvert at the intersection of his trail with the road, he had stepped into a muddy spot, sinking deeply into the slushy muck found there. Obviously, he had stopped on reaching the level surface of the road, placed his burden on the ground and had taken time to clean some of the mud from his lower body. It was this mud he had cleaned off himself that had drawn my eye. He then recovered his load and, stepping off the road, continued on to the south from that point. His stride was tremendous, nearly twice mine and I am six feet four inches tall and cursed by those that hike with me for my long strides. Yet, mine were as a child's when compared to his.

I had been talking softly to myself while working out this scenario as is my wont at such times. Finally, satisfied that I knew all there was to know about this, I turned to Frank. The specter that greeted me was absolutely hilarious and told me that he, too, had a good idea what had transpired here. He was as white as the snow itself. His mouth was opening and closing seemingly of its own volition, with no discernible effort on his part. Poor Frank looked very much like he was trying to articulate great words and thoughts but nothing was coming out. It was as if his mouth had been disconnected from the rest of his being and was left on its own. The poor guy looked very much like a goldfish without the bowl! I'm sorry now that I did it. At the time I had no real choice... it was all I could do... I laughed... Oh, how I did laugh. My sides hurt and my eyes ran with tears. It is a small wonder that I did not wet my pants, such was my laughter... I was totally beyond control. I have never in my life, before or since, seen such a sight.

When I finally calmed enough to control my mirth, I said, "Well, Bud, there is the source of our "woof" from earlier."

He just looked at me, his eyes wide. Finally, at long last, he found words and uttered a shaky, "I-is t-t-that what I t-think it is?"

"Yes," I responded with a grin, "It certainly is! Exciting, isn't it?"

With a look of sheer dread in his eyes, telling me he knew the answer to his question before he asked it, he said quietly, "Can we go now?"

I explained to him that he was free to go if he wished, but I was following those tracks. I simply wanted to learn more and this was the closest I had ever been to one of these creatures. I was not about to lose this opportunity. I had him close and had perfect tracking conditions. This was my best opportunity to get much closer to the creature we knew as Sasquatch. While this news did not seem to rank among the top ten things Frank wanted to hear just then, he was not about to go off by himself any time soon, so I was blessed with a partner in my quest… at least for the near future.

As I could see the direction of our creature's travel led back to the standing timber very near the point where our road entered that timber, I chose to follow the road to that point, intending to leave the road there and enter the dark timber on his track and see where it took us.

One positive aspect of this state of affairs, I suppose, was that at no time did I ever have to wonder where Frank was or to where he had wandered. I don't believe he was ever more that six feet from me and this was on wide open ground. It was to be expected then, I guess, that when I stopped suddenly he would ram me from behind.

This is precisely what happened when, just at the edge of the timber, I spotted a cougar track and stopped to point it out to Frank. In retrospect, I probably should not have further burdened his already over-taxed central nervous system with this rare and chance find. He did not take the news well. I never thought, however, it would cause the reaction that followed for the track was hours old… only still there because it was back under the protection of the canopy of snow-laden trees, safe there in a pocket away from the falling water. Already, nearly all of its mates were gone. A few were mere smudges in the snow, recognizable for what they were only because of the presence of that one clear print. As I started to explain to Frank, this was a younger cat, not yet full grown… probably just on its own away from its mother.

All this logic and calm thinking was lost on poor Frank. That the track was many hours old did not even register in his over-fevered mind. All he could do was begin muttering, "GET me out of here…Get me OUT of here… Get me out of HERE…" The volume rising with each iteration, of which there were at least ten.

At this point with a half-crazed man in my care, I had no choice but to abandon my search and see to Frank. Besides, by now, he was shouting at the top of his lung capacity, causing snow to fall from the branches of the trees around us. He had obviously reached the limit of his endurance. To continue further could have been dangerous, if not to me, at least to him.

"OK, Frank," I smiled calmingly at him, "let's head on out. You stay close behind me (some of the most superfluous instructions ever uttered… somewhat akin to 'take cover' on December Seventh in Pearl Harbor…) and we'll head back for breakfast. I think every critter within five miles knows exactly what and where we are after that outburst.

By this time it had rained sufficiently that the snow load on the brush and trees was beginning to slip off, allowing the weighted down shrubs to spring back to their "pre-snow" positions. Each time this happened there was the sound the snow falling off and the whoosh of the branch popping up. This is what triggered that involuntary reaction in poor, overwrought Frank… With each release, he would utter a loud, sharp yelp… somewhat like what one would hear upon inadvertently stepping on the tail of small dog… and he'd jump straight into the air while simultaneously executing a perfect three-hundred and sixty degree spin while airborne.

"At least," I smiled to myself and thought silently, reluctant to share negative thoughts concerning his demeanor with Frank just now, "I don't have to worry about anything sneaking up on us from behind!"

Ozette Lake Encounter

On the far west end of the Olympic Peninsula, very near the Pacific Ocean lays a small lake of some twenty four square miles (7700 acres). The local Makah Indians call it "Kahouk", or Large Lake. While its surface is some twenty nine feet above sea level, its bottom is over three hundred feet below that standard. We call it, today, Ozette. The setting for this eight mile by three mile lake is pristine and quite beautiful. From the west shore of the lake, it is only about a mile to the ocean beaches. Much of the lake is within Olympic National Park and is administered by the National Park Service.

The region to the south and the east of this lake are some of the finest timber growing lands in America. The soils are, on an index that ranges from a top of site one to a low of site five, mostly

Site Index One soils, the very best there is. These lands are owned, mostly, by companies in the business or growing and harvesting trees. When I was there and working, there were still vast tracts that had not been logged. I don't believe that could be said today. ITT Rayonier was one of the larger landowners in the area and I contracted to them often for road and bridge design work in this area of more than two hundred inches of annual precipitation.

This is a strange and eerie land that borders the river south of Lake Ozette. It seems we step back in time when we enter these ancient forests. The timber here is magnificent and it grows more rapidly than most realize. I have harvested forty year old timber that was more than four feet in diameter. It is an amazing place. As such, it is the home to some amazing beings as well.

Douglas Fir Forest

It was in this forest that I had my first encounter with sasquatch wherein I knew what I was seeing and where I could sit and observe for several minutes. This land is quite low elevation and tends to become swampy and difficult to traverse at times but it is most magnificent in so many ways. The story told here happened as described. My friend, Gary, would not discuss these creatures with me. I know he'd had encounters, but he refused to even discuss them. That being so, I never mentioned what happened this day. He also happened to be my wife's cousin and, knowing her family's attitude about these "wild men", perhaps had something to do with his reluctance.

Gary's family came from West Virginia. In fact, he was descended from the infamous Hatfield family that was involved in the famous feud there. Whatever the cause, he was not ever able to even discuss the subject with me.

Chapter 2

My Ozette Lake Family

By

Thom Cantrall

It was one of those gorgeous, late September days that make one glad he's alive and can be in God's creation. The sun was bright, if not overly warm. The air was so clear that every tree stood in stark relief, each an individual due to the early morning shower here in this land of many showers.

Gary and I had taken advantage of the weather to play hooky and see if we could bag a blue grouse or two, at least this was the excuse we gave ourselves. Actually, we both knew we were here mainly to enjoy the day and perhaps be able to spot some elk. To this end, we found ourselves near Lake Ozette on Washington's Olympic Peninsula.

Gary was a forester and log buyer for a hardwoods company and he managed their tree farm near Deep Creek on the very north tip of the Peninsula. My logging company was on an enforced fire danger shut down and had been for a few days. Dry weather and low humidily in the fall was a deadly combination in the woods. Low humidity coupled with unseasonably warm temperatures made for a dangerous fire situation. In fact, virtually every major forest fire in this area had occurred in September and October due to the described conditions. We both felt that the earlier precipitation coupled with higher humidity levels would probably allow the State Department of Natural Resources to lift the fire ban and we could return to work the next day. That made this day that much more precious.

Our perambulations found us, late in the morning, on a small no-name tributary to South Creek just a couple of miles south of Lake Ozette. We were stopped in a gravel pit there to glass the surrounding area for elk. Just to the east of this pit was a clearcut that was clearly visible to us in our elevated position and we judged the age of the brush growing there to be about perfect to feed elk and deer. We had been at this for twenty minutes or so when I spotted a deer feeding her way along the edge of the cutting unit. I told Gary where to look to find her as I watched her going about her own business. As he was trying to locate her, he suddenly jumped. "Look to the left of that big, burned stump west of the creek," he blurted out suddenly.

I switched from my twelve power binoculars to my eight power glasses because of the much wider field they present. They make it much easier to locate something in a wide area of search. I began a systematic search of the appropriate area of the clearcut trying to locate a large, burned stump in a sea of large, burned stumps. After a few minutes of search, I spotted what he saw… a very large bear ripping away in a rotten log there in search of some tasty grubs for his dinner. He was sleek and fat, obviously well conditioned for his soon to be enforced period of sleep. It is not every year that bears hibernate here. The weather is normally just too mild to require it for the most part and food is available to them all twelve months of the year. Some years, however, the snows hit this area hard and heavy and they will find themselves a warm, dry spot and sleep through the worst of the winter. This fellow was obviously well equipped for just such an eventuality, should it come. For the better part of an hour we watched him as he contentedly ripped the log to ribbons and feasted on the hordes within. Actually, bear were in season here and we both carried tags for the critters, but it just seemed a sacrilege to ruin such a beautiful outing by turning it into work. Add to this the fact that the spirit of the bear has special meaning to me, and I just was not about to disturb this fellow at this time.

As we kept an eye on him, we continued glassing the rest of the unit and eventually, as we knew would happen, a band of Roosevelt Elk made their way into the lush feed provided by the open aired unit. As is their wont, they did not come far into the open, but fed mainly along the edge, taking advantage of the brush growing even back into the standing timber because of the increased light levels now available there. As we watched, about sixty head were moving along slowly. That there were more in the timber and out of sight was obvious as well. There were cows and calves and the occasional young bull but no larger bulls as we felt there should be at this time of year. Late September is the onset of their breeding season and the rut brings out the big bulls. As we watched them, we heard a bugle sounding. It was the high screeching squeal and grunts of a large bull announcing his presence. We didn't know then if it was the warning call of the herd master or of another, challenger bull.

This is why we were here. The two blue grouse in the Bronco were the excuse, but this was the reason! We studied the layout very carefully and decided on an approach tactic. Our goal was to circle around to the far side of the herd and approach them from within the standing timber, thereby catching them between us and the open clearcut. Our wish was to get in close enough to get some good pictures but, just seeing them close up would be fun.

Our plan worked to a tee… we approached the feeding herd from within the standing old growth timber until we reached a point where we were actually inside the herd. I had elk all around me. I wasn't threatening them, just walking along with them slowly and talking to them with my cow call. They would look at me and mew softly or bark at their calf and return to feeding. We had not yet spotted the herd bull but had seen several smaller bulls… last year's calves mostly and while they were no longer nursing, they hung close to their mother. As we continued along with them, we knew there was a ninety degree angle ahead of us and they would then have to go into the open clear cut and we would be able to see the big bull. That was our thought and our plan. The actuality was somewhat different, however, as when they reached that corner, the bulk of the herd would not go into the open, but turned back and nearly ran over us in their escape. I saw one five point bull during this retreat and I thought the whole herd had escaped back past us when I heard a noise ahead of me. There was a small hill just before me, so I hurried up that hill and found myself at the very end of the standing timber. Directly in front of me were about thirty elk in a small depression. They were bottled up there as they could only exit it one at a time so had to wait on the one in front to leave before the next could make the short jump to get clear. I was in awe of this specter when I heard a bull bugle again and looked up the hill to my left… there he was… he was absolutely magnificent. His antlers were very high and sported eight points per side. When he laid his head back to bugle his anger again, his antler tips were nearly to his hips. He is, even until today, the largest Roosevelt Elk I have ever seen. Keeping in mind that a large Rocky Mountain Elk will weigh around eight hundred pounds, live weight, this bull would top fourteen hundred pounds easily. He was absolutely massive!

I watched as he continued to call his harem back together and slowly made his way out of that clearcut and back into the adjacent timber. For several minutes after he left my sight, I could hear him still bugling. He was still calling his cows back to him. I saw over a hundred elk in that herd. I don't know if he would breed them all, but I have no doubt but what he could if given the chance.

Roosevelt Elk in Timber

As the action died down, I knew that I had totally lost contact with Gary and should make my way back towards his Bronco so I chose a route and started out. I knew I was probably two miles from the truck, so decided to make a long swing through the timber, keeping to higher ground to avoid the chance of getting into one the myriad swamps that plague this area. Although

Area of Blown Down Timber

I knew it added distance to my trek, if it kept me out of the low stuff, I'd be quite happy.

As I moved down off the ridge I was walking, I spotted a break in the trees ahead of and below me. This happens fairly often in this area and it generally indicates an area too boggy to grow even cedar trees. Since this was on my route, I decided to get a bit closer to it to see what I was dealing with here. As I neared the opening, I realized it was not a bog or a swamp, but an area of blown down trees. It covered about six acres, perhaps a bit more and the blow down was pretty neatly arranged, looking for all the world like it had been fell and bucked. The trees lay side by side, only the upturned root wads on some of them showing that this was not the result of some logger run amok. Many of them were broken off from eight to thirty feet off the ground. The result was an open area in the midst of a pure stand of timber. This type of opening is normally caused by lightening striking one, usually the largest, tree in the vicinity and breaking it off. In an old growth stand, this creates a large opening and a heavy wind can cause a neighbor tree to be blown down into the open area. This action continues until an area as much as ten acres is created. In particularly heavy storms, entire regions can be blown flat. In 1921 such a storm devastated the Pacific Slope of the Olympic Peninsula and there are vast stretches of timber that started growing after this storm. It flattened an area about fifteen miles wide by a hundred miles long.

As I moved closer to the edge of this curious anomaly, I noticed two figures about fifty yards from the timber edge. They appeared to be feeding and the memory of the big bear we had watched earlier leaped to mind. I quickly found myself a spot where the wind was directly from them to me and I could sit quietly and observe them for a bit. As I raised my eight power binoculars, my twelve power glasses having been left in the truck, for a closer look the first thing that struck me was that these were definitely NOT bears. They were bipedal and they were bent over picking something off the downed log in front of them. I looked around me and noticed Oyster Mushrooms growing in profusion on these downed trees. It was obvious that this was what they were feeding on. Oyster Mushrooms are very tasty eaten right off the log and I envied them their feast.

For several minutes I watched them feed from my hidden spot at the edge of the timber. I had a perfect view of these great creatures as they diligently fed. Every minute or so, first one, then the other would rise up to its full height and survey the area around them. One, while standing was over a foot taller than the other. I was quite sure the larger was a male and the smaller a female until she turned toward me and I could see her fully frontal... that removed all doubt. She had very ample and evident breasts. Thus there was now no further question as to what I was witnessing. With great interest, I listened to their communication and marveled at what I was seeing. The male was especially magnificent. As I watched, he would examine everything in a three hundred and sixty degree circle. When he saw something unusual, he would stop and make an utterance to the female and she would hunker down low until she was almost indistinguishable from the logs. I wondered at this behavior until once as she rose, I could see she had a young one with her. It was quite small, in comparison, not even large enough to be seen above the downed logs until she lifted it. I immediately went back to my glasses and watched this strange family as they fed on the pure white Oyster Mushrooms.

Sasquatch

I was totally enthralled with what I was seeing here and felt no compulsion to leave anytime soon. Quietly, the family fed toward me until, at last, they were not more than twenty yards distant. Their conversation was remarkably clear at their closest point and I was in a kind of heaven as I watched my incongruous family and was hoping it would last a bit longer. I knew that, by this time, Gary would be back at the Bronco but he would not be worried about my tardiness as we had done this often and invariably every time one or the other of us got distracted and were late returning. The standard procedure in this case was to break out the Coleman stove we always carried and put on the coffee pot and heat up a can or two of whatever we had at hand.

As I watched quietly, wishing there were a way I could talk with my guests, I suddenly felt a breath of air on the back of my neck where, heretofore, the breeze had been in my face. Instantly, both heads snapped directly toward me. I knew they had made me! My idyll was over, so, with both the male and female trying to locate me in my hidden spot, I merely stood up and moved to where they could see me easily. I didn't want to frighten them and felt that if they knew what was there, they might not be so frightened. That they would now be leaving, I knew of a certainty, but they didn't have to be scared off. I spoke to them as they turned to leave. "Thank you for allowing me to see you," I said quietly. "Please be of good health with your family."

For a moment the pair stopped and both looked back at me as if they didn't know exactly what to make of me. Slowly they turned back in the way they were going to depart the area. When they reached the edge of the standing timber, the female moved on into dark timber with the youngster while the male stopped and turned slowly back toward me. He made no sound, nor did he gesture, but for a long moment he paused and looked directly at me, holding his gaze for just a bit before he nodded his head at me and turned and stepped into the timber following his mate and child.

I waited a bit longer in case he might return... although I knew he would not. I took a few moments to gather some of the delicious mushrooms and resumed my trek out of the woods. As I had expected, when I broke out of the timber and walked the last few yards to the Bronco, there was Gary with a hot cup of coffee and a lunch prepared of Chef Boyardee Ravioli and potato salad. To this I added a nice batch of the translucent, white Oyster Mushrooms and we enjoyed a pleasant repast there beside a tiny logging road in the middle of nowhere, each lost in his own thoughts. Gary tried a couple of times to get a conversation underway about the elk we had seen. I acknowledged that subject and even participated in a desultory manner, but, truth told, my mind was not on elk at this moment. I wanted to share what I had seen, but I knew Gary would not understand. He had no experience of this magnitude with these beings and he was certainly not going to believe my experience of spending time with a family.

I kept my counsel to myself about my family. But, I knew... I knew I had been the recipient of something very special. I knew these beings lived as we live... in a family group. They are devoted parents and care for their young as we would care for our own.

Thank you, Dear Family, for allowing me this privilege.

Dickey River Encounter

It is not by coincidence that these first three encounters took place within a very small geographical area. From the Hoko River where Frank and I found the tracks and heard the big fellow woof at us to this encounter on the upper end of the Dickey River is only a distance of a very few miles. It is, however, or was at the time of the incidents, a most primitive area. From an area just south of the Hoko River all the way to the Quillayute River near the town of forks stretched a nearly unbroken carpet of old growth timber. Few incursions had been made into this virgin forest and my perambulations into the region were, wherever I went, unprecedented. Simply put, I was the first, other than the rare hunter, to place foot onto these moss covered grounds. When I found a footprint on a ghost of a sandbar in a stream I could easily step across while miles from the nearest road, I KNEW that no one had been there to make that for me to find. I could not imagine any thought more ludicrous than thinking that some person would be in that far and remote place to create a false track in a place man had not trod in probably forever wand with no idea that anyone would again for ages.

This land was generally low elevation and subject to flooding when the rainy season was on. Since this area had an annual precipitation level of over two-hundred inches per year, that was pretty common. In fact, the day I was here to execute this contract was chosen because the water had just receded back within the banks of the Dickey River making access to the area I needed to measure finally possible.

This day stands out in my memory as one of the most significant days of my life as it concerns these beings. I learned things from this encounter that I still use today in calculating facts from data surrounding sasquatch. It was a revelation to me when I was able to calculate a ground pressure figure from the encounter. The tracks and trackway were interesting and watching this large male actually walking provided me with information I was unable to understand for many years until I learned what a mid-tarsal break actually was and what it meant in their pace. I watched his foot bend in the middle as he walked and was amazed by it but did not realize the importance of this for some years.

Chapter 3

Dickey River Encounter – Eye to Eye in the Timber

Who Is Sasquatch?

By

Thom Cantrall

Olympic Peninsula October, 1977

He was large, over eight feet tall and easily weighed six-hundred pounds… he was covered with long, dark hair. His massive head seemed to sit directly on his broad shoulders with little or no neck between. Oversized crystal-brown eyes surveyed me diligently as I stood transfixed. We watched each other at a range of less than ten feet with my mind cataloging all I was seeing even while my brain was telling me this was an impossibility. "This creature does not exist and anyone who states otherwise is either lying, perpetrating a hoax or is misidentifying what they are seeing," the so-called expert had said with authority.

Well, at this moment, deep in the swampy morass known as the Dickey River country, I wished fervently that I had this "expert" with me. I was over two miles from the nearest road and more than a mile from anything that could be considered even a ghost of a trail on a trek that no one other than my partner knew I was taking. My partner was nowhere within reach. He was probably sitting in my truck on the road eating my sandwich and drinking my coffee. One thing I knew for a fact. He was not going to be anywhere near this far from a soft seat and dry cab. I knew that no one was going to know anything of what I was seeing except I tell them.

Frame 71

I was here in the capacity of a Forest Engineer to do a bridge site survey for a local timber company who planned to build a road into this stretch of virgin timber. As that road crossed a salmon spawning stream, a hydraulics permit was going to be required to satisfy the requirements of the state in order to obtain the necessary permits to build this road. A bridge site survey involved measuring the size, width and depth of the stream as well as the soil types that are found. In short, everything that will go into the design and construction of the bridge and adjacent road was to be enumerated and recorded.

Sasquatch... Bigfoot... Swamp Ape... Yeti... they were the same creature and they did not exist. I was told so by experts... So, why was something that does not exist standing there watching me so intently? Didn't he know he was an imaginary creature... a myth? This was evidently not true, for he began to move slowly away from me. He was walking upright, just as I do. Perhaps I was misidentifying a bear? It doesn't seem likely that someone who had hunted and harvested bear would misidentify one from this range of ten to twelve feet, does it? He began a slow retreat towards the nearby timber and away from the berries he had been feeding on when I first spotted him. He did not totter in a lumbering fashion as a bear does when he walks on his hind legs but walked smoothly with a strange little hitch in his gait. In abject awe, I watched as he walked surely and directly to the heavy cover. As he went, he turned occasionally without stopping to assess my actions. He need not have worried for I was not moving from where I stood. It was as if I had taken root exactly there.

This encounter etched itself into my brain as I realized that this was probably what that expert called panic hysteria induced by some event in my childhood that caused me to hallucinate and think I actually was seeing what my eyes were recording... but, why wasn't that expert here? In fact, I wondered if that expert had ever been here... or any similar place anywhere in North America? Somehow, I don't really believe he had been.

When this creature of my imagination disappeared into the darkness of the timber I stood and watched as his image seemed to be burned into my retina. When a few minutes had passed and this after-image abated, I decided to see if I had actually been hallucinating and moved to where he had been eating berries. Unfortunately for my reasonable expert, the first things I saw were tracks... large, wide tracks there in the soft ground. Five toe prints were clearly visible and from heel to toe, the track measured seventeen inches in length and was approximately five and one-quarter inches in breadth across the ball of the foot. It tapered to a width of about two and three-quarters at the heel. I could see a series of his tracks between where I now stood and the timber across the way. The first thing that struck me was the length of the creature's stride.

I was a Forest Engineer and, as such, had taught myself to walk with a measured pace. I was able to measure long distances by pacing and be accurate within fifty feet in a mile. I had done so many times, often to the amazement of my partner. My pace, left foot then right was exactly five feet. I could maintain this pace accurately uphill and down.

Using my calibrated paces, I carefully measured the stride of my visitor and found his pace, from the heel of his right foot to the left and back to the heel of the right again was within two of my paces or over nine feet and below ten feet. As I had watched the creature walk away, I knew he had not been alarmed and was not running, but merely walking steadily on his way.

The last measurement I wanted was the depth of his footprint in the soft ground as it compared to my own. I knew that my shorter, narrower foot should penetrate deeper than his long, broad feet. To test my hypothesis, I removed my boots and socks and walked as he did over the same ground, just being careful not to obliterate his tracks. I was amazed that my foot did not sink deeper nor, interestingly enough, did his sink deeper than my own. In fact, we made very similar tracks separated only by size and the arch in my foot. Both showed the balls of our feet, five distinct toes, a marked arch in mine and a round heel. The only real difference was in the fact that he seemed to place his foot more evenly on the

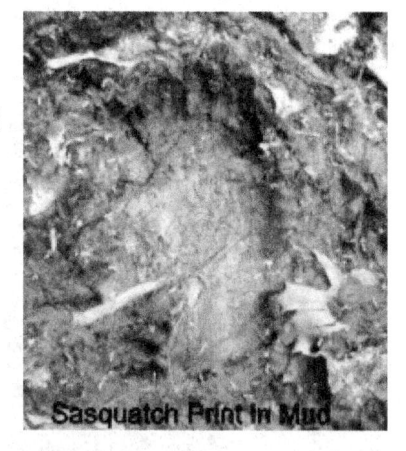

Sasquatch Print in Mud

North Fork of Dickey River

54

ground than did I, not rolling from heel to toe as I did in my paces.

As I sat and replaced my boots, it struck me to measure my foot and compare that to that of my non-existent visitor. When I did so, I made what was, to me, a startling discovery. I computed my foot to have covered approximately thirty square inches. And, since I weighed two-hundred-twenty-five pounds fully dressed at that time, I was exerting approximately seven point five pounds per square inch of pressure on the ground. When I measured my imaginary guest's footprint, I judged it to be approximately eighty square inches and, while I did not know exactly what he weighed, as he didn't seem prone to staying around while I found a set of scales, but I could estimate the weight of cattle quite accurately and I felt I could be just as accurate with this myth. When I divided my estimated weight of six-hundred pounds by the eighty square inches of his foot print, I came up with an identical seven point five pounds per square inch! No wonder we sank so nearly the same in the ground, we were exerting virtually the same pressure per square inch on it as we walked.

As this was before the days of small cameras that took excellent, high resolution photographs such as we have available to us today, and I was packing all my gear for many miles on my back, I had no camera available to me on this trek. Therefore, no pictures exist of this sighting nor of the trackway he left for me to decipher. I did, however, record all the measurement data from this trackway into my "Rite in the Rain" notebook for later review. I sketched the scene and indicated all the pertinent data found. It should be noted that until this day, I still use the figure of seven point five pounds per square inch to determine the approximate weight of these beings.

Armed with all this data, I continued on to complete my bridge site survey and began my hike back towards my truck…

When I arrived, I was right... my lunch had been ravaged and my partner was sleeping contentedly in his corner of the truck.

Del Norte Encounter

In the spring of 1978 I relocated from Washington's Olympic Peninsula, paradise in my eyes, to take a position with a timber company located on the Oregon-California border right on the coast. I came here because they needed someone to run their road building crew as they had purchased two major timber sales from the US Forest Service. In the interim from the time I was engaged for the position and actually reporting, the USFS had put a hold on two of our sales as they were involved in litigation with the ultra environmental groups that control most of California.

Taking a Break 'Cause the Bull Buck Showed Up

I was asked if I would mind working with the Bull-Buck (person who runs the timber cutting crews) as his number two man until the USFS made those sales available to us. Had I understood all this would entail, I would probably not have been so excited by it...

Many mornings I left the mill yard at two am to get to our jobs out of Orleans by six am to meet my cutters. By the time I had them where they needed to be and had met with the USFS administrator to solve the dilemma of the day with them, I would be headed back home in order to get a late dinner and be in bed by nine pm. It was a grueling schedule as I had to do this three to four times a week.

Our sale administrator and NEVER done this before. He knew nothing about his job nor anything about logging... much less timber cutting. He had been a log scaler for the USFS for a great number of years and when the opening came up for a sale administrator, he bid for it and because he had the seniority, he got it. As part of the government senility system, knowledge of the job is of no importance. The only criterium that matters is longevity. This fellow had the "Monte says" syndrome. His boss's name was Monte so everything he said was said was prefaced with "Monte says..." It was totally frustrating to have to explain every little nuance to him in minute detail as to why what they wanted would not work or why you could not turn a hundred foot tower on an inside curve with a radius of fifty feet. I had to explain it to him and he would then have to go to Monte for approval. I was beyond help by the time I was shut of this particular Piss-Fir Willie.

The one thing positive that did come from this debacle was this experience. In location, Blue Creek is the next drainage west of the much more famous Bluff Creek of Patterson-Gimlin fame. The biome is very similar. It is not the lush green of the coast a few more miles to the west but the douglas fir was huge in the areas where it was not subject to direct sunlight. North facing ridges grew fantastic timber while the south facing ridges ran more to Manzanita brush until one got into the canyon bottoms where shade aided in the growth of trees where sufficient sun was interspersed. It was a very critical biome that had been visited with wild fires time and time again which created the huge trees with burned out stumps that I found here.

  It was this constant conflict with the USFS over minute changes that caused me to leave this position permanently. I had learned an animosity in dealing with the USFS while operating my own company for so many years. I, and every other contractor who did work for them, knew that you always had to increase your bid on any job for them by at least twenty-five percent because they were going to keep you sitting idle twenty-five percent of your contract time. If a decision could not be made by the individual on the ground, then it was never going to be a part of this job. By the time anything went "back to the office" for clarification and then got back to us, we were well beyond the problem at hand… most often because we had waited until the bird dog left at four pm and then just did what needed doing without his knowledge.

  But, that is the politics of dealing with any bureaucracy and simply is what it is. This encounter was probably the crowning glory of my life until I met my teacher. It was so obvious that the male of this trio was trying so hard to communicate. I watched him as he tried to slow down his delivery… increase the volume… speak softer… articulate oh so carefully… we tried them all without achieving positive results.

  I used this, in a form, as the prologue to my first book, "Ghosts of Ruby Ridge" and used this area of burned out stumps in the body of that book because they affected me so profoundly. I pray you feel the spirit of this night as much as I felt it there with them.

Chapter 4

Del Norte Encounter – A Family Affair

By

Thom Cantrall

In the spring of 1978 I worked for a timber company located at the California-Oregon border on Highway 101. My job required that I drive from the mill yard inland to our logging jobs west of Orleans, CA. To get there, I had to drive a huge circular route.

Cat Road

Leaving the yard, I drove south on US 101 for just over sixty miles through the coastal Redwood groves to the Bald Hills Road just north of Orick, CA. I followed the BaldHills Road for about thirty-six miles to Weitchipec, CA and Highway 96 where I turned north for approximately fourteen miles. At Orleans, I turned back to the west and drove for about twenty five miles to our job sites. Since much of this was driven on gravel roads, the trip required four to five hours to complete, depending on the amount of traffic on the highway. I was required to make this trip an average of three times a week.

There existed at that time a road that ran directly from the small town of Gasquet, CA, up the south fork of the Smith River and past Doctor Rocks and on into Orleans. This direct route shortened my trip to about ninety minutes and was known as the Gasquet-Orleans Road, or more familiarly, the G-O Road. It was paved on both ends, but there was, in the middle, from east of Blue Creek to the west of Doctor Rocks a stretch that had never been constructed beyond a bulldozed trace through the timber. The U.S. Forest Service had plans to finish this road, but was being fought vigorously by the lunatic fringe preservationists who pretty much control California, it would seem. The final result being that, looking at a current map of that area, it shows that road still not being completed. In fact, a Wilderness Area is now shown to enclose the Blue Creek crossing and the area these tracks were first found.

Much of this primitive section of the road was at sufficient elevation that winter snows drifted deep and kept the track closed until early summer, at least, under normal circumstances. This particular spring, the

60

shortcut being so important to us, we hauled a D-7 Caterpillar as far in the west end of the road as we could before the snow stopped us.

There, we unloaded the Cat and let him clear snow across the ten miles or so until he broke out of it on the east end. We used a rubber tired road grader to clear what drifts were amassed on the paved section east of the primitive road. At the beginning of the pavement, we reloaded the Cat back onto its trailer and hauled it on to our road construction site.

With the G-O Road open to Four Wheel Drive vehicles, our crews could leave home two hours before time to be at work on Monday morning, work their time, spend the week in Orleans at a logging camp we'd set up there and return home after work on Friday. If there were something sufficiently important to do at home, they COULD make the trip in midweek, though this was frowned on. On the week in question, I had meetings scheduled with the U.S. Forest Service Sale Administrator on Thursday to set where the roads into the next unit would be located. I then had a conference with our road construction boss set for Friday morning. I determined to drive over on Thursday, have my USFS meeting, spend the night at our camp in Orleans, meet the road boss on Friday and drive home Friday afternoon.

I was meeting with government workers on Thursday so I knew I could sleep in a bit longer as they would not leave their office in Orleans any sooner than 8:30 am.

Since I knew I would be staying over, the pack I always carried with me in my truck in case of emergency was especially plush that Thursday morning as I pulled out of the mill yard at 7 am. The sun was well above the eastern rim when I reached the snow line on the G-O Road. That I was the only vehicle to cross this morning was evident in the icy slush that was on the road in various places.

Trackway

61

I had traveled about a half mile from the point the snow began and was on a slight uphill grade traveling west to east when I spotted tracks in the snow. The tracks came from the north, dropped down into a shallow swale that opened onto the road in a very muddy stretch. They continued on south, up the slight bank on the south side of the road and disappeared into the distance.

My first thought on seeing the tracks was that a bear, just out of his winter's sleep had been on a trip of exploration, probably for his morning meal. I am always interested in locating sizable critters, and especially since there were no cub tracks I could see, it would most likely be a lone boar, I stopped short of where the tracks crossed in the mud of the road to measure this bear. As I walked up to the tracks, my jaw dropped like a rock! There in the muddy slush was not the bear tracks I expected to see, but a very large, very human shaped foot print… not just one, but a whole series of them.

For several moments I just stared! Bare, humanoid foot prints that measured just over eighteen inches in length with a stride that I, at six feet, four inches could not begin to emulate. For me, a full stride, left and right is exactly five feet in length. I've measured it time and again in my capacity as a forester. The stride on this creature was well over eight feet in length! That was an awesome stride! My first inclination, after regaining mobility, was to follow them to see where they led, and, hopefully, what was making them.

I had but little time to devote to this. A multi-million dollar logging operation could not be left to falter because I wanted to chase a Sasquatch. I did flag the spot well, so I could find it easily on my return trip. I knew I could be done by noon on Friday because I did not have to wait on the USFS and could meet the road boss on the job at six am.

Tower Side

Noon Friday found me in my little truck, climbing the last grade out of Blue Creek Canyon that led to the crossing… not that I was anxious or anything. When I reached my markers, I found a secluded spot without much snow where I could park my truck out of sight of the road. I knew the cutting crew, the logging crews and the road building crews would be passing through here tonight and, knowing that most knew my truck, I did not want them to know what I was about doing here.

When I was ready to travel, I set out on the now day old tracks with little hope of catching up with this particular creature, but I had to follow. Down the ridge we went in the snow. Within a half mile, we broke out of the timber onto a sunny, south-facing slope that was clear of snow except in the very shaded areas. Every few hundred yards there would be a patch of snow varying in size from a few feet across to some that probably covered more than an acre. Although it was not difficult tracking in the bare trail that varied from damp to muddy, these snow fields served to let me know I was still on the same individual.

Beautiful Trackway in Snow

Very late in the day, when I felt I had hiked about eight or nine miles from the G-O Road, hunger was beginning to rear its demanding head so I decided to look for a good campsite, enjoy my dinner and take a little time to explore my immediate area before dark spread its tentacles and drove me back into camp. The area I was in was populated with stands of magnificent old-growth Douglas Fir of huge proportions. Some of these were more than seven feet in diameter and it was obvious that they had survived many, many fires. Between the stands, especially on the south facing slopes, the scars of those fires were very evident.

When I dropped down onto a flat gravel bar adjacent to a beautiful, clear running stream, I thought I had probably found my campsite and when I noticed that several of the huge old behemoths had their trunks burned out, leaving a warm, dry, cave-like den, I determined that I was at home for the night. This had everything I normally look for in a campsite, level ground, cover from possible lightening storms that the current increasing clouds could certainly deliver, and abundant fresh, clean water.

The only disconcerting thing about my campsite was a rather putrid smell that wafted through from time to time and, in searching the den burned from the tree trunk, there were a large number of long, black hairs lodged in the bark and wood. I thought I had probably found a bear's winter den and, since they were out and doing now, they would not mind sharing quarters with me, since I was determined I would not be there when next they needed it for hibernation. This area, as I have described it here was the model for the second Sasquatch camp in my book, "Ghosts of Ruby Ridge".

Burned Out Tree

The first thing I did after getting out from under my pack was to hike up the stream for a couple of hundred yards, checking closely for dead critters lying in the water.

The coming night was just beginning its tenure when I heard from the timber the most god-awful, gut wrenching, piercing, high, ululating cry. It was absolutely stunning and bone chilling to hear. I had, at the time, absolutely no idea what could be singing that song and I wasn't really sure I wanted to know. I had heard descriptions of the call of the Sasquatch, but, believe me, no description I had ever heard even began to prepare me for the reality of it. The first call went on, varying in pitch and modulation for what seemed like minutes, but

Hollowed Out Tree

which was probably between thirty and forty-five seconds. It then ended by fading away in volume to zero.

I was sitting by my fire, completely at attention, eyes and ears under full strain to learn more when it began again though not in the same place. Where the first call was to the south, this call was from the northwest. Again, the same high ululations, almost a warbling sound followed by a steady tone only to be varied again. This time I was able to be a bit more clinical about it as I was not quite so in awe of the sound in and of itself.

I timed this scream at twenty-five seconds until it again faded away. When the calls ceased, there was not a sound to be heard from any source save two. The bubbling of the small creek which was wholly unimpressed with the nocturnal display I had just witnessed was one sound. The other was the thumping of my heart in my chest. I judged the calls to be just up the ridge from my lair, not over two hundred yards away from me.

After these two calls, nothing more was forthcoming. I built my fire up slightly so that it afforded more light. When about two hours had elapsed with no more contact, I noticed a shadow flicker across one of the openings to my den. A moment later, another shadow appeared. They were not really close to my tree, but just at the edge of the light cast by my fire. I quickly searched my pack for the flashlight I always carry there. Unfortunately, when I found it, I could not get it to work. My pack seldom leaves my truck so that I always have it in an emergency. Normally, I remove the batteries from the hand light and store them separately in a plastic baggie to prevent what I had just discovered. Evidently, at some prior time, I had broken my own rule.

Without artificial light, I was relegated to making the most of the light my little fire afforded. By sitting near the opening with my fire at my back, I was able to see my "guests". There were three of them that I watched most of the night. Evidently, I had unwittingly commandeered their den and they did not appear overly pleased with the prospect of sharing it with me. At any rate, they were with me all night long, a night that lasted, I might add, approximately one hundred and seventy seven hours.

The relative sizes of the three individuals led me to know that I was dealing with a family group here. The male was huge. I would estimate his height, judging from the brush he was near from time to time to have been over nine feet. The female, for obviously she was, was not nearly so tall. She appeared to be in the seven and a half foot range and the youngster was not over five feet tall. The male had the most massive and powerful upper body on an individual I had ever seen in my life. His body tapered to a waist that I would have been proud of and his legs looked like massive pistons. His musculature was phenomenal. The female was built strongly, with massive upper body strength but with a thicker waist and prominent and evident breasts. Junior was a miniature of this father. All were dark in color and the two adults had high, crested heads while Junior had a very round head without evidence of a sagittal crest.

For literally hours I was in their presence. Never once did I feel threatened nor did I ever really feel a need that I should leave my nest in any great hurry. During these hours, Papa tried diligently to talk to me. I could readily hear them "talking" to one another. I didn't know it at the time, but the utterances I was hearing were not uncommonly made and would become the subject of expert analysis in years to come. Ron Morehead had, in prior years to this, recorded very similar vocalizations to those I was hearing now and would in the future, team with the cryptolinguist, Scott Nelson, to begin the research into their language. That I knew they were speaking language was clear. Papa would speak for long periods looking directly at me from a range of less than twenty feet and often within ten feet. He would articulate very carefully in trying to help me understand him. It was so frustrating for us because nothing was working to help us understand. Just as you and I might slow down our speech and pronounce so clearly to help someone who does not speak English to understand us, so did he in his language. Where we might speak louder in the hopes that would help, so did he…. I was most amazed at what was happening before me this night.

Unfortunately, it ended all too soon. I was tiring as the night waned. I thought about staying with this group somehow and, perhaps, living with them, but I didn't know how to effect this. I had a son coming soon and I knew I owed him more than this, but it hurt so deeply to end this wonderful night. I have never been able to emulate it but I would hope I might be able to do this before my tenure on earth is due.

Towards morning, I dozed in short catnaps that were often interrupted by the sounds of woofs and yips that I heard from outside my nest! Somewhere near dawn, these sounds stopped and, my fire built up to last a bit more and I slept soundly for a time.

When I woke, light covered the land, my fire was burned down to coals and the only sounds to be heard were those common to the mountains in the daylight hours.

On emerging from my retreat, the first things I saw were myriad tracks. From these tracks, I discerned that there where, indeed, three separate creatures of three separate size classes. As soon as I had completed my breakfast and morning ablutions, I hoisted my pack and my butt and hied out of there and back to the road and my waiting truck. I have always wanted to go back in there and check that place out, but I left that job and that area within a month of this incident, and have not been back in that area for any period of time since this occurrence.

Tree bower

This incident is factual and is reported here exactly as it occurred. The memory has remained bright in my mind though more than thirty years have elapsed since that night.

Butterfield Canyon, Utah Encounter

    This day started out go just be a day outside. I hate being cooped up in a city. The only redeeming factor they seem to have is that they are such wonderful places to escape! Lynn and I often did this kind of thing... usually ticking off our wives when we didn't return at the prescribed time. I am not really sure what was so important that we needed to be inside an apartment by a particular time. Nothing was going to happen there anyway. So we went... and returned when the situation dictated.

    This is one such time. Had not this big fellow arrived when he did, we probably would have been able to keep the commitment to return on their time... but he did... and the rest became history. It had little effect on me as I was already sleeping on the floor while visiting... Poor Lynn may have suffered.

Chapter 5

Utah Encounter
By
Thom Cantrall

It was late summer and the antlers were hard now, if still velvet covered... The beginning of what the Ute Indians called the Moon of the Crying Elk, though there was no evidence of them bugling yet. In fact, the bulls still seemed to be running in bachelor bands as we saw three different such bands on this day.

It was in 1993 as I remember. My son in law had graduated from BYU the previous spring and had taken a position with a firm in the Salt Lake Valley. He was living in the Halliday, UT area. Cabin fever was hard upon us when we decided a voyage of exploration might be in order. Why we chose the west side of the valley that day, I have no idea. Perhaps those in charge (our wives) had placed a time limit on us for our excursion as our previous outing that had begun at Provo Canyon had lasted until the wee dark, single digit hours of the following morning... they were

Mule Deer in Velvet

prone to do that in those times for some perverse reason. I can't say that those restrictions ever really limited our outings to any degree although that lack of limiting discipline might have been the cause of the two of us fixing our own dinners from time to time. Today was to prove no exception.

We traveled west across the valley on one of the streets that was a direct route... 4100 S. I think, in looking at the map... we intersected the main road on the west side of the valley near a powder company (Hercules Powder, perhaps?). From there we journeyed south with the idea of finding a likely looking canyon that would allow us a hike into its bowels. Several times in this area, we saw elk in the distance, including two bachelor bands, one of which held two exceptional bulls, the larger of which would have probably been a 320 class bull or better. Since all of these appeared in direct correlation to an ample supply of "No Trespassing" signs, we assumed that Utah elk had the ability to, if not actually read, at least to recognize signs! On one occasion we spotted a band of six Mule Deer bucks feeding on some green feed very near an old farmstead. As this area was lacking those hated signs, we decided to put a stalk on them, eventually closing to within bow range of the bucks. There were two in this band that I would have happily brought home. Both were 4X4s with extremely high and wide racks though still fully in the velvet yet.

We traveled on south past the great mine at Bingham Canyon to a point just south of there. There is a road there that turns up Butterfield Canyon and we decided we had time to explore this area before curfew... well... at least not MUCH beyond curfew... How far up that canyon we were, I, at this late date, cannot recall. I guesstimate about five miles. In glassing to the south, slightly past a large pile of rock that I guessed to be tailings, we spotted another bachelor band of elk. We could not see how many or what quality these animals were, but, since there were no signs to tell us otherwise, we decided to check it out. I parked my old Bronco, Widowmaker, out of sight of the road and we headed up a small draw…

It was not far up this gulch that we came to a split and my son in law took the south arm while I took the north. About twenty minutes later as I lay glassing a clearing that lay before me and between me and Lynn, I saw a shape in the small trees that defined the far side of the clearing. I lay very quietly in my place and watched closely, fully expecting a muley doe to come out of that brush as the glimpse I had had been far too dark to be an elk. Very suddenly, the dark shape darted from the cover it had been in while hiding from me and entered an even thicker, darker copse beyond. It was a large being… running on its hind legs… at least seven feet tall and possibly as much as seven and a half feet. It was massively built and very dark. Its arms were long, as they had to be and it was very cautious of me. I had no problem understanding what it was, though I was mildly surprised to see it here. I wondered why it had revealed itself to me with that sudden dash when I heard brush cracking and Lynn emerged from the thick cover. He had followed his branch to where it had become too difficult to continue, since we were not really hunting, but just out curing a fever, and had decided to move in my direction.

Author in ASAT Camo Glassing Far Ridge

As had happened so many times in our years of hunting together, his move was timed impeccably and worked to our mutual benefit. I did spend a bit of time looking for tracks which were evident only as disturbances of the leaves on the ground. I also took the opportunity to unobtrusively estimate the height of the small trees he'd been standing amidst while hiding from me to better determine a height estimate.

We then retraced our steps to ol' Widowmaker and headed on home… first stopping at a Burger King along the way to preclude the "punishment" we had waiting for us on arrival.

Revelations Encounter

For many years I spent almost every waking moment either in the deep woods or wishing I was in the deep woods. I lived in the country near the small town of Sequim (Pronounced skwim) on the very north end of Washington's Olympic Peninsula. It was an idyllic place... a veritable Garden for me. I fished her streams for trout and steelhead and her saltwater for salmon, halibut and cod. I took crab, clams, shrimp and oysters in profusion. I hunted her wild country for deer, elk and, occasionally, bear. In all the years I lived there I purchased meat so rarely that I generally forgot that section of the supermarket even existed. In addition, the timberlands of the peninsula provided my living. I owned and operated a small gyppo logging company, buying and selling most of the timber I logged myself.

Having attained a Bachelor of Science degree from University of Washington in Logging Engineering, I contracted from time to time with the large timber companies to do work for them. This generally consisted of establishing property lines, cruising timber, fitting a road to the land or, most commonly, doing a bridge site survey. When crossing a stream used by anadromous fish, those that migrate from the sea like salmon and steelhead, there must be a hydraulics permit obtained to insure the safety and integrity of the stream. This permit required a detailed diagram of the proposed structure and of the stream and streambed. This included items like flow rate, depth and width of the stream at peak flow, and soil types in the area of the bridge. It was an extensive and detailed project to provide this data and I was called on to do many of them. During the course of these surveys, I was alone in the woods making measurements, reading instruments or writing copious notes. So many times, our large, hairy friends came in so close probably just to see what I was about. I learned to listen for them and even identified several of their calls.

One call, one which I was not to identify for approximately another thirty five years was a simple up-down, two note whistle. I heard that several hundred times in those woods and I simply attributed it to a bird I had not yet identified. It was not until the Oregon Sasquatch Symposium where I heard Ron Morehead's "Sierra Sounds" audio for the first time that it crashed through the shield of my mind and told me I'd been hearing them telling one another of my presence all that time.

This essay tells of one of my forays into the empty country of the Olympic Peninsula. Every word herein contained is true. This happened just as outlined here. There was more that occurred, but I am not able to share that at this point. I beg the reader to understand that there is more to come from this day and to simply accept that I am under constraints here.

Goodman Creek, where this incident occurred is just off the Pacific Ocean about equidistant between the Hoh and Bogachiel Rivers on the far west end of the peninsula. The spot where I was set up here is only a very few miles off the Pacific Ocean.

Chapter 6

Revelations of Novelty

By

Thom Cantrall

It was one of those days in mid November that is typical on Washington's Olympic Peninsula... WET... It was, however a wet that was not only endured but was revered because it was elk season and that meant an opportunity to spend time in the wilds alone. It would have seemed a dreary day to anyone not an elk hunter. The skies were not pouring water, but merely weeping softly much like a young girl who has just learned her best friend talked on the phone to the boy SHE liked... real enough, but not in profusion... and certainly nothing when compared to what could be.

In those days, unlike today, you could hunt with a lesser weapon in a season primarily designed for a more modern weapon. Hence, I was afield with my trusty recurve bow and solid aluminum arrows in the modern firearms season. Although that would seemingly put me at a great disadvantage, it was not ever about the actual taking of an elk. Yes, that would be nice and it would certainly ensure my winter's food supply, but the main thing was, I was here in God's creation enjoying what He has made and that was the real reason. I was not at a disadvantage in method anymore than I normally would be with a bow because by this, the third weekend of the season, you never saw another hunter off the road. The real hunters had their tags filled and were home. The only ones here this weekend were those like me who sought the experience and those who managed to fill a tag about once in ten years when an animal was dumb enough to walk across a road somewhere in front of them… or, like me, had a late season antlerless permit and could afford to be very selective in what they harvested. Once off the roads and into the timber, I saw no other man.

On this day I had chosen a spot on the south fork of Goodman Creek just a few miles from the Pacific Ocean for my evening hunt. I had hiked the mile or so into the area I wanted in mid afternoon and found myself a perfect outlook over the creek bottom. The creek bottom here was relatively flat with a series of beaver ponds as far as I could see both upstream and downstream. That those beaver ponds were also full of spawning salmon, I also knew for I had had one for dinner the evening before. In all, it was a beautiful scene laid out before me. I was at the edge of a primal stand of mixed Douglas fir and western hemlock that was many hundreds of years old. Before me and just lower was the creek bottom that was mostly grasses and sedges with occasional small trees growing singly or in small copses next to the ubiquitous ponds. Beyond the stream, the land rose slowly to the opposite stand of timber duplicating that in which I was secreted. I was here simply because the elk fed here in profusion. There was a shrub growing here that they loved and it brought them in regularly and in numbers.

I did not have a tag for antlerless at this time, so it would have to be a bull with visible antlers, and the odds of any bull still being in a herd this late in the season were very slim, but, as I said, it was not about that. I loved being in this beautiful spot on earth and having the opportunity to spend this time just enjoying what may come. I was excited about the prospects of the evening and could barely contain myself or the anticipation building within me. My skin fairly tingled with the thoughts of what might be.

I had been in my hiding place for barely fifteen minutes when the first critter came into view on the trail that traveled in fits and starts between me and the nearest ponds and roughly paralleled the stream. With a furtive glance behind himself, the most beautiful ice-blue coyote I had ever seen sneaked quietly beside the pond directly below me... When he'd covered about half the distance from the beginning of the opening to the end of it, he stopped suddenly and his head came up, his body on full alert. Without so much as a quiver, he stood rigid, watching downstream in the direction he had been headed when, suddenly, he wheeled one-hundred-eighty degrees and tore back up the trail he'd just descended. All caution had been thrown to the wind in his obvious need for escape and I watched downstream, wondering what could have bothered him so much.

Canis Lupus

In moments, I was rewarded for my patience with the sight of something I had recently been told by the department of game on my reporting a prior sighting... up that narrow path along the creek trotted two very large, very alive timber wolves (Canis lupus)! Silently I watched as they moved steadily upstream or west to east in direction. They looked neither left nor right, but maintained a steady, loping, pace up the trail. Where are those fools who laughed at me last time, came as an unbidden thought as I watched the two gray ghosts quickly cross my line of vision and disappear into the standing timber further east of me. "Wow," I said aloud to no one I could see, "that was worth the price of admission for sure!" It was about this time that I remembered that I had my camera along for just such a circumstance and never even thought to use it... oh well... maybe next time! Each of these dogs would have weighed well over a hundred pounds and possibly as much as a hundred twenty pounds. They were magnificent as I watched them transit across my field of view from left to right. I was armed and there were no strictures at that time against harvesting wolves in this area as, as I was told rather forcefully, they didn't exist in western Washington. I could have, but

I certainly had no idea of doing so. They were simply magnificent in their carriage and mien. I was privileged to have been allowed this glimpse into their lives and I cherished it then as I cherish it today, more than forty years later.

I have been told that we as people are not good witnesses because we don't remember correctly. It has been posited that as time passes, we recall an event and then change our view of it in our memories and then later recall the new version and again change that view… to this, I can say a resounding "No WAY"… perhaps this is true of the mundane events… even perhaps the anniversary party where uncle John got so drunk… but when this type of occasion occurs, at least in my mind, it is burned into place. I can close my eyes today and see that scene as clearly as that November day so many years ago. Every tree is vivid in its place. The water in the beaver ponds reflecting the images of the huge timber to the north of me… and those two creatures… the brightness and shine in their eyes… carriage of their tails… the alertness of their ears and the beauty of the scene. I can certainly believe it is as Ayla and Jondalar may have witnessed it some twenty-five thousand years ago… it is that primeval… that ancient… that breathtaking!

I sat for several minutes after their departure wondering if I could even discuss this with any of my friends. The people I hung with were not known for their attentiveness nor for their supportiveness in issues of which they were not personally involved. My mind was well involved in this dilemma and I was only paying attention to my view in a cursory manner. My attention was still with what I had witnessed and not with what was to come. It took a bit of extra time for the fact that I was hearing a cow elk and her calf talking to register in my mind.

As the fact of this became more prominent in my consciousness, I brought my attention back to the present and began to search for that which was arresting my attention. The soft mewing of the well grown calf was answered by the sharp bark of warning from the cow as the sounds neared my position. It was nearly twenty minutes later that the pair appeared on that same trail that had been the subject of so much activity this evening. I was in for a mild surprise when I found she had not one calf, but twins. They appeared to be a male and a female as one had the small beginnings of antlers and the other did not. I watched quietly as they slowly fed their way along this pathway and wondered why they were seemingly alone here. It is not common for a lone cow to be with her calves and no others and this was making me wonder what I was overlooking when I heard another cow further to my right and on up the trail where this family group was headed… it then became evident as several more animals emerged from the standing timber and began feeding along the edge of the beaver ponds to the east of me. She was not alone, she was merely the further most west of the band. Although several elk were in range of me, none, save this bull calf before me, were legal bulls so I merely watched them feed along and listened to their music as they talked one with another as they passed.

Anxiously I was hoping another animal, a bull, this time, might be following that cow with her twins and I remained silent and deadly in my lair for quite some time past their final departure. The air was quiet. A soft rain fell… one of those rains in which you could spend all afternoon outside and never really get wet. I was so happy in my choice for having come here and was contemplating calling an end to my idyll when I heard water splashing a bit downstream. In listening closely, I thought it might be in the last pond before the stream entered the standing timber at the extreme west limit of this special vale. As I listened, it sounded like children playing in the water, something that could not be in this cold, wet late fall weather. I even lifted my binoculars to attempt to see more clearly that which I was hearing but to no avail. There was naught but to wait. Time might tell… After all, I knew that often large bulls will splash and cavort in water to some end and I was anxious that this might be the case here. Pending this eventuality, my senses were heightened. My

Goodman Creek Beaver Dam

nerves were set on a hair trigger edge while awaiting what was to be. That the noise was moving up the stream was becoming evident as I listened with every sense at its highest level.

Night comes early at this time of year and I knew my time was growing short. Normally by this time I would be starting out in order to not be caught in the timber after dark but I HAD to see what was making this racket. It now sounded like two animals thrashing the water, but I reasoned that this sound could be made by a bull walking in the water and thrashing it with his antlers. Whatever is was, this was not a small bull. This had to be a mature bull elk to effect this set of circumstances. Not so patiently I waited as the noise moved up stream from one pond to the next until, at last, I could see movement through the screen of brush and limbs in the pond just down from my position. Amazingly, what I could see appeared very dark in coloration, not the buff tan I had expected from an elk. Perhaps I was seeing his head and mane, I thought, both being very black on the Roosevelt Elk that populated this biome.

Time stood still as the shapes behind the trees moved inexorably toward the opening east of the tree line that now covered them. There had to be two animals as the splashing sounds were distinctly separate and I could see the one behind the screen of brush being quite still while a bit away, the thrashing of the water continued… The next thing I noticed was in the vicinity of the dam that created the next beaver pond to the east just at the head of the pond being beaten to a froth was that there were salmon jumping the dam to escape upstream into the upper reaches. Now, I wondered, what would make salmon feel a need to run from elk?

It was at this time that the first of my "elk" stepped from the shade of the trees… and showed me anything BUT an elk. My first thought was that this was a bear fishing for the plentiful salmon in the stream but then he stood up. This was no bear! He was, firstly, huge! I had no real way at that time to accurately measure his height, but I judged him to be over seven feet in height and weighing more than six hundred pounds. That he was male was evident and he was, except for the full covering of hair and his size, a man was also apparent. I watched fascinated as this large being kicked his feet and struck the water with his hands to create the splashing I had been hearing. I was not immediately aware of why he was doing this, but I could not ignore that it had some design to it.

While I watched this large male another, slightly smaller version of this same image stepped further into my view in the pond. That the first was obviously male I knew. That this one was just as obviously female, I also knew. Her breasts were evident and her difference in body morphology was also evident. Where he had huge shoulders, a barrel chest and a waist that seemed no larger than my own, she had shoulders and chest that were more subdued and a much larger, thicker midsection and buttocks. The differences when viewed side by side were profound. I was in heaven as I watched this scene unfold before me when, suddenly, the purpose of this exercise became abundantly clear. With a sudden motion, the male reached down quickly into the water and came up with a wriggling salmon. He quickly, while holding it with one hand firmly around its gills, grabbed the fish's head with the other and twisted like he was unscrewing the lid off a catsup bottle. The effect was the same as on that bottle… the fish lost his head as surely as the bottle lost its cap. The fellow then handed it to one I'd not seen prior who was following behind. It appeared to be an adolescent male as it was noticeably smaller in stature but was a miniature of the large male. Soon the female made a similar motion with similar results, but she dropped hers back into the water behind her.

That puzzled me so I watched more closely to see what was really happening here. In the next few minutes, in the waning light, the two adults caught approximately twenty fish, most of them as they neared the wall of the dam above this pool. Of those fish, they returned probably twelve to the water while keeping eight. I was at a total loss here until it dawned on me that they were squeezing the abdomen of the fish as they held it in their hand, using their forearm against their side. Those they returned had no eggs issue forth… all of those they kept had eggs. They were determining if the fish they'd caught were spawned out and, hence, in some stage of dying, or if they were still fresh, viable salmon. Just as I would never have kept a spawned out fish from a stream, neither would they. I used to, in this era, get spawned fish from the hatchery and we would use it for smoking but it was a very tedious job as we had to be so very careful with the meat so as to not transfer possible bacteria from the decaying meat to ourselves... over half of that given us by the hatchery was used as garden fertilizer, nothing more. It seems these large fellows knew the same song.

It was fully dark by the time they had moved upstream enough that I could no longer see or hear them. I don't know if they stopped their activity when darkness overtook us, but I could no longer hear them, so perhaps they did. I cautiously rose from my nest among the trees above the trail and began my egress back to my waiting truck. I had a good hand light with me as well as a headlamp I affixed by a strap to my forehead before leaving my blind. It was the work of but a few minutes to make my way to the trail that had been the mother of so many interesting things this day and it was another hour or a bit more to make my way through the standing timber on a ghost of a path so hard to see even with the portable lights I carried. When I finally arrived at my truck, I searched through my work pack and found my Rite in the Rain notebook and recorded all that had happened to me this wonderful day. That I would never be able to tell anyone, I knew, but I also knew I had to have record of it for myself. Too few minutes later, I was back on Highway 101 headed north and back to the town of Forks where I would have my dinner and reflect on what I learned today... reflections that would bring a silent and enigmatic smile to lips sealed to the events... until today, some forty years later when someone asked...

Todd Neiss Encounter

Sgt. Todd Neiss

Bigfoot witness-turned-researcher, Todd M. Neiss has been an active investigator for more than eighteen years. Born and raised in the Pacific Northwest, he grew up hearing of these legendary creatures, alternately known as Bigfoot or Sasquatch, but gave it little credibility beyond that of Native American lore or a good old-fashioned campfire tale designed to frighten young campers. All of that changed for Todd in the spring of 1993. As a Sergeant in the Army's 1249th Combat Engineer Battalion, he came face to face with, not one but, three of the elusive beasts in the temperate rain forest of Oregon's Coast Range while conducting high-explosives training. His sighting was independently corroborated by three fellow soldiers who also witnessed the creatures.

Since that fateful day, Todd has conducted numerous investigations including several long-term expeditions in the Coastal, Cascade and Blue Mountain Ranges of Oregon & Washington, as well as Northern California. Todd believes that, in the tradition of Jane Goodall and Diane Fossey, the best way to obtain credible evidence of the existence of these fascinating creatures is to infiltrate a small research team into the heart of prime Bigfoot habitat for an extended period of time. Ideally this should be for forty five to sixty day rotations. In doing so, he hopes to acclimatize the creatures to their presence and eventually overcome their inherent apprehension of humans.

It is his opinion that these creatures possess a relatively high IQ in comparison to recognized great apes. Todd's current theory focuses on that presumed intelligence which he believes fosters an irresistible sense of curiosity...a curiosity which he intends to exploit. By presenting a variety of baits as well as an array of unconventional, non-threatening lures within a pre-designated area, he hopes to successfully collect irrefutable evidence of these creatures existence.

"It is my goal to entice these animals by presenting a non-threatening posture and piquing their curiosity, thereby luring them into a specified area where irrefutable evidence can then be obtained," says Sgt. Neiss. Once the creatures are officially recognized, his ultimate goal is to establish a management program to ensure their perpetual existence for future generations to appreciate.

Over the years, his research has garnered him international attention. He has been the subject of numerous documentaries and TV programs from the US and Canada to the UK and Germany. He has been featured on such programs as: "Unsolved Mysteries," "Strange But True," "Encounters," "To the Ends of the Earth," and both the Discovery Channel and Travel Channel." He has also appeared on a number of radio & TV talk shows, local newscasts and newspapers. He is currently wrapping up work with director Blake "Buck" Eckard on a documentary (four years in the making) for Stonehill Pictures.

He has been a featured speaker at colleges and universities, and a regular contributor to the *Annual Sasquatch Symposium* series held in British Columbia, Canada. Todd has instructed classes on

Bigfoot for the Audubon Society as well as the Campfire Boys & Girls Society; the former involving both classroom and field work and culminating with an overnight working camp in the "Dark Divide" of Washington State's Gifford Pinchot National Forest. He was a featured speaker at the 14th Annual Bigfoot Conference held in Newcomerstown, Ohio in April of 2002. Todd has investigated a multiple sighting in Oregon's Coast Range, as well as the infamous Ape Canyon (Mt. Saint Helens, Washington), the Blue Mountains of Eastern Oregon, and numerous forays into the Mt. Hood National Wilderness.

Todd currently resides about 40 miles due east of Portland, on the western flanks of the Cascade Mountain Range, in the small town of Sandy, Oregon. This provides an ideal location from which to conduct his research and respond quickly to reports in northwest Oregon and southwest Washington areas. He encourages anyone with a legitimate sighting to contact him immediately. He promises to treat every report confidentially as well as professionally. Todd can be reached by e-mail at americanprimate@aol.com.

Todd is the founder and host of the annual "Beachfoot" conference held every summer on the Oregon Coast. The event started out in 2008 as a simple, invitation-only gathering of friends and fellow researchers in a private meadow nestled within the temperate rain forest near Lincoln City. Beachfoot 2009 saw the numbers swell to just over 50 participants and included notable researchers.

Chapter 7

My Bigfoot Encounter

By Todd M. Neiss

Few moments in life have such a dramatic impact on a person's life that it qualifies as an "epiphany" - a moment where one's concept of reality itself is utterly and permanently altered. Such a moment happened to me one sunny day in early spring of 1993. April third was a day which is, and will always be, seared in my mind as if it were yesterday. Even after more than eighteen years, the specifics of that event should illustrate the impact it had on my life.

It should be noted that, with regard to even the possibility of these creatures existing, I was beyond skeptical. Simply put, I had relegated these beasts to the realm of Native American legend or merely a classic campfire tale to frighten young, gullible children. I rarely watched sci-fi programs and never had read a single book on the subject. Ironically, as fate would have it, I would later become the topic of many such programs and books.

As a sergeant in Charlie Company (1249th Combat Engineers), it was business as usual as we headed up into the dense temperate rain forest of the Coast Range in Northwestern Oregon. The mission of Combat Engineering can be boiled down to two words: "mobility" and "counter-mobility". In other words, ensure our troops overcome any obstacles (man-made or natural) and deny the enemy passage (or route them to where we want them) by placing obstacles in their path. On that particular day, our mission was to conduct training on private timberland near Saddle Mountain just east of the coastal resort town of Seaside. We would be executing demolitions operations at three rock quarries. Each site had a unique battle scenario to accomplish.

At the first site, we practiced "cutting charges". This is where we would use plastic explosives (composition four or C4) to shear steel I-beams like a hot knife through butter in an effort to simulate dropping a bridge. Simultaneously, we also cut a five foot diameter Douglas fir tree in half by wrapping a belt of C4 around it. Both charges were a resounding success! One sheared steel I-beam looked like an exploded cartoon cigar and a whole lot of bark dust.

Exploding Landmine

The second site held a "complex obstacle" consisting of a field of surface-laid anti-tank mines followed by a triple-strand concertina wire fence. We were to clear a vehicle lane through both. In addition, we were tasked to construct a field-expedient 'claymore' anti-personnel mine out of a #10 coffee can with improvised shrapnel. After securing the area and checking for subterranean mines, we strung a 'ring-main' (a circuit of detonation or DET cord) through the mine field. DET cord looks similar to fuse cord with the exception that it contains a tremendously explosive compound (PETN or Pentrite) which burns at a consistent rate of 8,000 meters per second. It is said that if you could string a line of DET cord from LA to New York, it would take approximately 14 minutes to get to the other end! It is essentially used to synchronize several explosive devices to detonate virtually simultaneously. You definitely do NOT want to confuse DET cord with fuse cord! In any event, my squad set about placing C4 charges next to each of the anti-tank mines and tying them into the ring-main while another squad began fashioning a field-expedient (read: homemade) version of a 'Bangalore Torpedo' to breach the razor-wire obstacle. Normally a Bangalore Torpedo is essentially a 3" plastic pipe filed with C4. Sections of this pipe are generally fitted together to form a pipe long enough to breach the entire obstacle. In this case we had to sandwich C4 between sections of U-channel fence pickets then wrap them together with duct tape (same effect). The homemade 'Bangalore' was then tied into the ring-main. Lastly we constructed our field-expedient claymore mine by poking a hole in the bottom of a #10 can and inserted a blasting cap. Next we lined the bottom of the can with about 2.5lbs of C4 then covered it with three layers of cardboard for wadding. Finally, we loaded the can with rocks, bolts, nuts and anything else that would ruin the 'enemy's' day. The can was then buried into a hillside (pointed towards the enemy) and angled about 12 degrees off the ground then it too was tied into the ring-main.

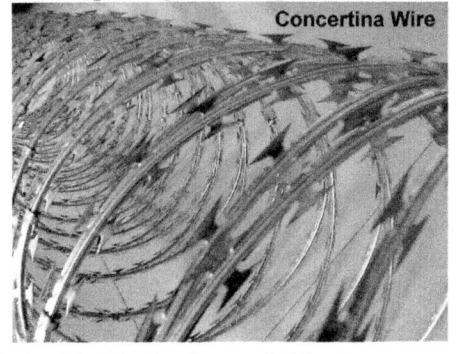

    While I had not yet seen these creatures, there was a brief incident which, in retrospect, made me think they may have seen us. While I was directing my squad to emplace their charges next to the anti-tank mines, there was a rather loud, crescendoing 'WHOOOOP!' that emanated from the west end of the mine field. At that moment, I was bent over placing my own charge. Upon hearing this somewhat shrill noise, I immediately stood up and sought out the perpetrator as we were under orders to practice noise discipline during the exercise (in case the 'enemy' were nearby). As I glanced around the mine field, I was surprised to find all of my men still busily preparing their changes and not, as I suspected, goofing off. I shrugged my shoulders and went back to work. In hindsight, it seemed to me that the WHOOOOP sound had came from farther back in the tree line. But that made no sense as everyone was present and accounted for.

Once all of the charges were set and the area was cleared, I yelled "FIRE IN THE HOLE!," then pulled the dual-primed M-60 fuse igniters. The fuses hissed and began snaking their way towards their primary charge while we mounted up and began to convoy down to the safety staging area to await the 'report' of the explosion a short eight minutes away.

At that moment, we developed radio problems. The field commander could not reach the base commander back at Camp Rilea. I was tasked to take my HMMVE ('Humvee') up to the top of a nearby hill, where we had a 'two-niner-two' radio relay station set up, to see what the problem was. Upon arrival, they had already repaired the relay, so I decided to watch the impending explosion from that vantage point. Even from two miles away, the sight of 200lbs of C4 detonating is an awesome sight. The huge flash was followed by an even bigger black cloud which began to build into a mushroom cloud. Simultaneously you could see the trees in the immediate vicinity shudder in succession as a shock wave rolled across the forest below in a perfect concentric ring. Finally, about two seconds later, we heard the BOOM! ...Another resounding success.

The third and last training area was situated in yet another gravel quarry on a hillside that overlooked the second blast site. Here our mission was to emplace a 'cratering charge'. As the name implies, this type of operation involves the making of a rather large hole. Generally this is done to sever a road thus denying the enemy use thereof. To the uninitiated, an explosion is an explosion. To those of us who deal in the science of explosives, there are very distinct differences based upon the target, its composition, type of explosive (dynamite, C3, C4, ammonium nitrate, PETN, TNT, RDX, etc.), amount of explosive, its placement, shape of the charge, tamping, etc. Whereas C4 produces a super-hot/fast explosion, ammonium nitrate (essentially refined chicken or pig manure) soaked in diesel fuel for several hours, results in a 'slow' concussive blast. Properly placed and tamped, it will effortlessly relocate a generous section of real estate. It should be noted that, absent a standard issue shaped charge, we had the 'heavy junk' (read: heavy equipment) section pre-dig a starter hole with a backhoe. After emplacing several bags of diesel-soaked ammonium nitrate into the aforementioned hole, we (read: privates) filled it in and tap danced on it to tamp (pack) the charge. Once again, the area was cleared, and I initiated the dual-primed M-60 fuse igniters. I took my place in the waiting convoy and, per S.O.P., we began the descent down to the safety staging area.

Being a squad leader, I had the privilege of having my own Hummer, complete with a driver and an A (alternate) driver. Ours was the second vehicle of a five-vehicle convoy (2 Humvees up front, 2 covered troop carriers called "deuce and a halfs" and the Commander's Humvee in the rear). I took up a position behind the driver's seat and, as we were descending the narrow winding road down toward the staging area, I had the opportunity to enjoy the scenery. As an avid hunter, it is just second nature to me to spot for wildlife. As it was a rare sunny day in April, I had my window unzipped for a better view. Rounding a corner, I had a good view of the rock quarry where we had done our second blast at less than an hour earlier. Standing right out in the open, in the middle of the gravel pit, were three, jet-black, bipedal creatures. They stood in line (shoulder to shoulder) staring directly at our convoy as it descended the hillside across from them. Between us was a ravine populated with eight to twelve year-old Douglas Fir and hemlock 'reprod'. At a distance of several hundred meters, I could not make out facial features or gender, but there was no doubt what I was looking at were not humans. Had these creatures been standing in front of a backdrop of trees, I most likely would not have seen them at all. But in this case, there stood three dark black figures contrasted against a light grey cliff of basalt on a bright sunny day.

In the middle stood, what I assumed to be, the alpha male of the group; as it towered a full head above the two creatures that flanked it. I would estimate it to have stood approximately nine feet high, with the flanking creatures approaching seven feet in height. Their silhouette was unique in that their heads sat directly on their shoulders with no visible neck. They all displayed broad, square shoulders and barreled chests which tapered down to a svelte waistline, unlike the creature seen in the Patterson-Gimlin film of 1967 (for the record, I am of the impression that the PG creature was either pregnant or had recently been so; accounting for her girth). The arms of these beings hung well past their knees. In the case of the two flanking creatures, they were exhibiting a swaying motion (rocking side-to-side) as the larger creature stood as still as a statue. Bear in mind that, all the while I was staring at the creatures, we were bounding down a dirt road with the occasional hedge of blackberry and Scotch bloom obscuring my view. That being said, I had approximately twenty-five seconds of viewing time.

At this point most people ask me, "Didn't anybody else see them?" "Why didn't you say something to your driver(s)? or "Why didn't stop your vehicle?

The answer is that…
    I assumed I alone had seen them;

    I was still in shock and disbelief;

    my jeep didn't have a radio to call for a stop

    and even if it did, we had a rather large BOMB ticking off behind us!

Once the vehicle rounded a sharp corner, I knew I had seen the last of them. I fell back into my seat with a mixture of shock for what

I had witnessed and an odd sense of depression. It's a hard to explain what goes through ones mind in such a moment. Fate had somehow came together to create a once-in-a-lifetime moment that was now lost as suddenly as it had found.

My head began to swim with questions.

"Oh my God! They DO exist! And not just a solitary beast, but a group of them!
How could they exist and not be 'discovered'?

Having extensively hunting this area, how could they exist and I not have seen them, or signs of them, before? Some hunter!

What do I do now?"

I felt the sudden urge to tell someone, but who? I had seen something scientifically, if not historically, important and SOMEONE should be notified! There must be an authority that NEEDS this information!

I began to make a mental checklist.

The US Fish & Wildlife? No.

The Forest Service? No.

The zoo? No.

The police? HELL NO!

Then WHO?!?! And better yet, who would believe me anyway?

In the final analysis, I reluctantly decided (like most people do) to keep my mouth shut. Here I was a family man, a vice-president of a shipping company, and a Non-Commissioned Officer in the Army National Guard. I had worked long and hard for my reputation. And yet, with one simple sentence, "I saw Bigfoot!", I could throw it all away. Nope. If I knew what was good for me, I could never tell a soul.

Thus is the curse of the Bigfoot: living with the burden of the truth. A truth so absolutely incredible that merely suggesting that you 'might' have seen something that 'may' have been a Bigfoot will cause people to question your very sanity and even destroy your reputation. Great! I have jokingly suggested that I should start a Bigfoot Support Group for those afflicted with 'the curse'. One thing I can say, from years of interviewing other eyewitnesses, that there is something therapeutic in sharing such mutual experiences.

Arriving at the staging area, I immediately jogged back up the road in a futile effort to get one more look at these amazing creatures. Unfortunately there was a knoll which blocked my view of the gravel pit. Again an odd sense of depression swept over me. I felt a genuine sense of loss that was difficult to explain.

My activity hadn't gone unnoticed. Suddenly I heard footsteps heading my way. Then a voice yelled out, "Hey Neiss!" I turned and saw Sgt Jeff Martin heading my direction. As he approached me, he looked over his shoulder to see if he was being followed. Satisfied that we were alone, he said something that I will never forget. He took a long drag off of his cigarette, exhaled, looked me squarely in the eyes and said, "I don't suppose you saw what I saw back at the second blast site?" It was more of a statement than a question. I could tell from the look in his eyes that he knew something. I felt overwhelmed at the possibility but decided to err on the side of caution. I replied, "I don't know Jeff, what did YOU see?" Once again he looked left then right to make certain of our privacy and then stated rather matter-of-factly, "I saw three, huge, hair-covered, for lack of a better word 'BIGFEET'.

Trying to contain my excitement I hissed, "Yesssss! I saw them too!"
I was overwhelmed with sense of utter relief! I wasn't alone!! It

wasn't that I needed validation of what I had seen. Corroboration could not have altered the truth, but it sure felt good. It felt somehow liberating. At that we began to compare notes.

Fate was busy that day. What were the odds that I ever would have seen them in the first place? ...The odds that someone else did (independently) as well? The odds that they would have even imagined that I had shared their experience and even in so considering would have had the courage to ask me? I can only guess that observing me looking in the direction of the quarry and straining to get a glimpse of 'something' was enough to pique his curiosity. Thank God! And finally, what were the odds that we were the only two witnesses? As I would come to later learn, we weren't!

That evening, we had an 'open post' so I decided to stay the night at the home of my friend (and Platoon Sergeant) Don Braden and his wife Lena. After debating whether to tell them my story, I reluctantly opened up (few beers didn't hurt either). Being my first 'Bigfoot confession', I found out the hard way that even your best friends can be hard to convert. After some initial ridicule, I had to settle for a bit of patronizing sympathy. This wasn't to be my last bout with ridicule.

Fate wasn't quite done yet. At the next Guard drill, Lena Braden and the other wives and girlfriends of soldiers were conducting a bake sale in the foyer of the armory at Camp Rilea as was their custom. I, on the other hand, was split-training in Portland some 100 miles east. Lena was in the middle of sharing my 'Bigfoot confession' with the other ladies when two soldiers entered the building. As they passed the bake sale table, they happened to overhear Lena describing my encounter, then froze in their tracks. They turned to her and asked her to repeat the story then both confessed to having seen the creatures as well!

A lot has transpired since that fateful day. It has been my personal mission to provide irrefutable evidence of these amazing creatures existence in an effort to gain their recognition and, if need be, play a role in their protection. I have spent countless days and nights conducting field research (including six major expeditions) throughout the Pacific Northwest, California and one foray into the

Mazetzal Wilderness of Arizona. My quest has given me the privilege to meet and/or work with some of the top researchers in the field: Peter Byrne, the late Rene Dahinden, John Green, the late Professor Grover Krantz (a.k.a. 'The Four Horseman'), Bob Gimlin, Loren Coleman, Don Keating, the late Richard Greenwell, Larry Lund, Dr. Jeff Meldrum, Cliff Crook, Chris Murphy, Dan Perez, Joe Beelart, Ray Crowe, Dr. Wolf Henner Farenbach, Rick Noll, Thom Powell, Cliff Olson and the late Fred Bradshaw to name just a few. I thank them all for their generous insight, advice and companionship. Over the past 14 years, I have had the honor of appearing on more than nineteen television programs, several radio talk shows, and given speeches at numerous symposiums and colleges. Bigfooting will always be a part of my life and I look forward to many more adventures in the future

[some who read this might ask themselves why I included so much seemingly extraneous information. The answer is two--fold. First of all...since that fateful day in 1993, I had never written down a fully detailed account of my sighting. Secondly...while the events of that day are still quite fresh in my mind, it has been fifteen years and there is no guarantee that I will always be so cognitive. Lastly...in my years of interviewing eye-witnesses, I have always regarded the more detailed accounts as the most credible. I want to know what preceded and what followed the encounter. If someone allegedly took a photo or video of a Bigfoot, then surely there would be photos or video that preceded it (if not, then it raises "red flags" for me). So this is the reason that I felt it necessary to be as concise as I can be. I appreciate your indulgence

Missy Hutzler

One day, not so long ago, a very special friend from the Mountain State of West Virginia approached me to inquire about the possibility of me writing the story of her encounter. If you know me, you know there are two things I can never say "no" to… one is a chance to write about the sasquatch people and the other is a pretty girl. Now, here was a double header… a pretty girl asking me to write about a sasquatch encounter she had as a youth.

    This is that story. It is presented exactly as Missy told it to me except I added my style to the telling. The salient points are exactly as they were presented to me and what resulted is a most endearing and poignant story of the true nature of this being we call sasquatch. This encounter is one more nail in his coffin of humanity. This big fellow did what any competent man would do… he reduced the danger so the children in his domain could play safely… could any of us have done more?

Chapter 8

One Summer's Day

By
Missy Hultzer
as told to
Thom Cantrall

It was a warm late summer day in Martinsburg, West Virginia. A soft breeze was doing its best to keep the temperatures bearable as my cousin and I took our game of "Spiderman" to the trees of the apple orchard that lay behind his house.

The welcome zephyr rustled the leaves gently as we sprang from tree to tree while pretending to be the "Spidey-Siblings" practicing our awesome Spidey powers and web-slinging abilities. While I was but eight years old and my cousin a year less, we were diligently proving to each other that our characters could do all the powerful things that Spiderman could do. When we had ventured no more than ten or fifteen trees from his backyard, we stopped to plan our next adventure. It was during this planning break that we heard something break.

The sound filtered to us through the trees from the boundary. We, thinking it was Uncle coming to scold us, fell silent and watched closely, peering through the leaves, for him to appear. We were not allowed to play in the orchard for many reasons, primary of which was the farmer didn't want us there. Of course there were snakes and other dangers for children. Being children, however, we ignored those reasons as simply being excessive worry on a grown up's part. What we saw next was to remain burned into my memory forever…

As we each stared into the other's eyes in fear and disbelief, a very wide, very hairy man walked out from the grove and into the path that lead directly between the rows of trees where we hid. It was late summer, that time between picking time and fall, as there were no apples on the trees and only a few scattered on the ground beneath us. We didn't speak…we didn't move… we didn't even but barely venture to take a breath… I watched as this man… this creature, turned toward us. I was afraid… so very afraid, but, strangely, I felt I was safe because I was sure he could not see me. After all, I had been quiet, and he gave no sign…not even a glance up, to indicate that he saw us there. As the giant approached us he kept his eyes to the ground. It looked almost as if he was trying to avoid eye contact. I was surely was not trying to institute it either, for I simply wanted to know who this was.

I knew that none of my cousin's neighbors were this big nor were any of them in long reddish brown hair as this fellow was. The being was big but not as tall as one would think him to be for his bulk. He was built quite square really, with three or four inches of a reddish brown hair blowing gently in the wind. The only skin that I could see was on his hands and it was of a tan hue. It was much like the skin of grandfather, weathered and tanned from a lifetime of farming. As he approached our tree fairly quickly it seemed his purpose and movements were fluent. From the moment he broke through the tree line he was obviously on a mission and, just as obviously, I just had no idea what that mission was.

In a single motion, he approached us and squatted in front of our tree. Quickly, with hands that were but a blur, he reached out and snatched a snake from the base of my tree. He did this with a hand that appeared to have a thumb on either side of 4 fingers. With a casual nonchalance, he tossed it behind him like it was just a stick and of no consequence. He then stood and continued down the path without as much as a notice in our direction. Remarkably, as he got further away, he just kind of faded out of our sight. He didn't actually disappear but merely blended into the background as he increased our separation.

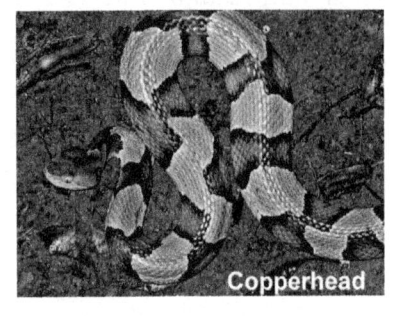

In quiet astonishment, my cousin and I simply stared one to the other for a few moments. Almost as if each knew the thoughts of the other, we both dropped from the tree and made a bee-line directly for the house. In an impossibly loud voice and at a rate only two excited children can attain we related to my aunt precisely what had transpired. To our great chagrin, she laughed it off as some crazy kid's story. To add even further to the distress we were suffering, everyone laughed at us and made sport of us for days. As so often happens in such cases, we vowed to never speak of it again. To this day he refuses to discuss it, and I, have not spoken of it to anyone until very recently. As it happened, I noticed another cousin with a bigfoot badge on her profile picture on Facebook. I took a major chance, my heart in my hand and told her of my experience. She then introduced me to my first bigfoot group on facebook.

Today the old orchard has gone the way of too many such orchards, having succumbed to a housing development but I have many thoughts on bigfoot and I feel my experience has changed me and changed my life. Perhaps I will write a novel on this one day, but for now, I will merely continue my search for my next encounter. Why? The answer to that question is simple… They are real, and they call to me. That is why. Is any other reason needed?

Kathi Blount

There are few people in the world who are as special to me as this lady is. I met Kathi for the first time at the Oregon Sasquatch Symposium in Eugene, OR in 2010. After the conference we opened a dialog concerning our large friends. Kathi related her experiences with a clan on the coast of Washington (the state, not the city) that she had a good interchange with for quite a long period. This attractive young woman is amazing in her understanding of these special beings and I hold hopes for great things from her in the future. In actuality, this chapter will contain two separate events.

First, we will listen to her description of her coastal clan. That will be followed by an occasion in which I was involved. In October of 2010, Kathi called me to tell me about a rough time she and her research partner, Cristy had received at the hand of a very large fellow on the Clackamas River in western Oregon. He was intimidating and had caused them to spend the night in their truck in fear of what could happen. We made plans for me to join them the ensuing weekend and this chronicle will tell that story.

101

Chapter 9

Kathi Blount Report

By Kathi Blount

As Told To

Thom Cantrall

There are some reports that bear being told just as they come. Here is one such report. The only thing I have done to Kathi's words here are some minor grammatical corrections and to make her narrative to all in general and not just to me alone. This makes it, technically, a paraphrase and not a quote. In essence, however, these are her words telling her account of the circumstances surrounding events that happened in 2009 and later in 2010.

The night was 4th of July

2009. Cristy, my friend, had come with me and we had another couple come up and camp in our spot also. They had two Pit Bulls with them that liked to bark A LOT. So I assumed that we wouldn't hear or see anything that night. Everyone went into Long Beach (Washington) to watch the fireworks on the beach. Cristy wanted to come back after only an hour at the beach. We arrived back at camp to find ourselves alone there. I think it freaked her out to tell you the truth because she immediately wanted a bon fire. I had to keep telling her to quit adding firewood to the fire!

    The other couple arrived back at camp and the woman took the dogs and went into their tent and went to sleep. I imagine that's what happened because I never heard a peep out of her or the dogs again that night. Brandon went into their tent and Cristy and I went into mine about midnight. I had placed my tent away from the others and nearer the lake. I was sitting in the dark with the tent door open while smoking a cigarette listening to the sounds of the night... or should I say "lack of sounds of the night" because the frogs weren't singing or anything.

    Suddenly, Brandon walked past the tent and towards the lake with something in his hand. I didn't want to talk to him, and the way he was sneaking past the tent told me that he didn't know that I was sitting there with the door open either. Suddenly, a bunch of lights and sound were flying out of his hand. He had lit one of those rocket fireworks that shoot out like twelve rockets. I'm not sure what they are called, but they are common. My first thought was "Oh how pretty they look over the lake".... my second thought was "What a dumbass, why are you doing that?"

    Just as soon as the fireworks were finished shooting out of the tube and the world faded back to black, something SMACKED a tree right outside of camp with enough force to actually reverberate in my head! I've never seen a man move as fast as Brandon did that night!! He ran back into camp and actually did a baseball slide towards my tent. I saw him sliding in towards the door of my tent

Western Lake

and I pushed him back out of it the minute he entered. He was babbling "Oh My God, did you hear that", while I was saying "Hey, you pissed them off now you deal with them!"

This was coming from a man who had just told me that morning that he and my husband were talking and neither one of them believed in sasquatch! Well, I'm pretty sure he believed then. At that point all I could see in the dark were the whites of his perfectly round eyes. By this time we could hear tree knocks all around us and all around the lake. It was so awesome!

That's when I remembered Cristy laying in the sleeping bag and I tried to get her to wake up and be a part of it, but she just kept repeating, "I'm asleep and I don't want to wake up". So, I think she may have just been too afraid to want to witness it. Here, she was talking to me, and the night up at Memaloose we weren't talking while it was happening. Brandon ran for his tent and I never heard another sound from that tent at all. I stayed up for another couple of hours and never heard any walking or any other

'bout as steep as it gets!

noises at all. They are so stealthy! I love that about them.

A few weekends later I was alone up there again. I was sitting by the fire roasting hot dogs about midnight. I was bored, in case one is wondering why I was roasting hot dogs at midnight. I figured if I cooked lunch the night before I wouldn't need to waste firewood the next day. I was sitting with my back to the bushes that separated my camp from the forest. I heard a distinct whistle behind me to my right and decided to ignore it. I was surprisingly peaceful about it. (I didn't think of this until later though) Only a minute or so later, I heard this whistle again. It was RIGHT behind me on my left this time. I said out loud, "Oh, are you hungry?. I'm going to put some hot dogs on a plate for you". I then put some dogs into buns and put them on a flimsy paper plate. I then said, "I'm going to place these on this pole and you can just reach thru the bushes and take them and no one will see you. In fact, I'll go into my tent and allow you to eat in peace. I'll come back out and get the plate."

That's what I did. I didn't have a camera with me and frankly, probably wouldn't have taken pictures even if I had one. The "pole" was actually just a cement post that they had garbage cans chained to at one time. The paper plate hung over all sides of this "pole" roughly two inches. I knew that no small animal or bear could take the dogs off of the plate without knocking the plate off the post. I went into my tent, read a book for awhile and then went back out to see if they were done. The plate was still balanced on top of the cement, the dogs and buns were gone and I felt fantastic!! I did notice a small piece of the bottom corner of one of the buns lying on the ground next to the pole. It wasn't bitten, it was torn off! And no, I didn't take it. I left it so the baby possum I saw in camp earlier could share their meal with them.

Kathi has other stories that have come from this beautiful, seaside lake. Unfortunately, limited resources and lack of reliable transportation prevented her from visiting this spot as often as she wished was possible but what was learned here put her in a grand position to advance her studies at a location nearer her Oregon home.

Unfortunately, all did not proceed well with Kathi at this new location. I was in this area and it is criminal what has been done there. It is the very near the Portland, Oregon metropolitan area and, as such is easily accessible to any who feel the need or compulsion to dump litter in otherwise pristine environs. It was into this condition that Kathi and Cristy inserted themselves in very early October, 2010.

Again, in Kathi's words…

We went in peacefully the first time… we just set up camp and played native music on the CD player. We sat on the boulders overlooking the ravine and played our recorders. I'm just learning and Cristy plays Gospel's on hers, so it's always mellow and soulful. We left a sea gull feather and two white rocks…. one from each of us as a gift on the boulder. We fed the chipmunks and went back into camp at nightfall. It was so nice watching the sunset and playing our recorders. I played the harmonica for awhile, but I don't really know any songs yet. I just play from the heart still.

Here's a strange part. While sitting in camp Cristy said that her back had started to hurt and she wanted to go and sit in the truck. I was surprised by this but just said that I would go sit with her for awhile. It really was more comfortable than sitting on the ground.

Remember we were going camp at the lake but after packing in we realized that there were just way too many people there for us to make any kind of bond or start our trust with our Forest People. So we drove further up the mountain and started following spur roads to see where they led. We found a great clearing overlooking the ravine with a great path down into it and set up camp there instead. We were sitting in the truck and Cristy was laying back in the driver's seat and I was sitting in the passenger seat with my eyes looking forward out the windshield. I don't remember if I closed my eyes or not but suddenly I had this vision of three very large beings walking through camp. I whipped my head around to look out the side window towards the tent and the trail. I didn't see anything in camp (obviously), but right then my heart started pounding in my chest. I became lightheaded and, in fact, felt like passing out. I actually saw dark spots in front of my eyes. I closed my eyes and tried to talk myself out of freaking myself out like this. I thought maybe I was just getting scared for no reason. No, scratch that, I was terrified suddenly! I didn't want to look stupid in front of Cristy so I never mentioned any of this to her. In fact, I never once turned and looked at her. Suddenly, I found words coming out of my mouth... "We have to go.... I have to go... NO, we are staying... No, we have to get out of here!"

At this same time, I realized that I couldn't move any part of my body and my legs felt like cement blocks. I think that's what scared the hell out of me. Right after I whipped my head around to look out the side window and then straightened my head back to look out the windshield, I became paralyzed! This lasted a few minutes at least and like I said, I never mentioned this to Cristy until I began babbling. While I was babbling she never made a sound. As soon as the paralysis ended, my legs felt like they were "floating" into the air!

Kathy Blount

Just as I was analyzing why I felt like this, Cristy said in a very small voice, "My legs feel so light now".

I immediately heard myself say, in as small a voice, "Mine are floating". She said "NO, they feel light".

I said, "You describe it as light, I describe it as floating, what just happened to you?"

She told me that she heard me talking, but couldn't move nor reply at that moment! The same things that I was feeling! But we never fed each other the information until after the experience, so we didn't cause each other's reactions. Something happened up there that I don't understand.

We sat and talked about it for about five minutes and we noticed that the fire was dying down and decided to put more wood on and get the sleeping bags out of the tent so we could bed down in the truck for the night. By the time we talked ourselves into getting out of the truck and checked out the camp for awhile, and yes, honked the horn a few dozens times (not proud of this fact, but we were terrified!), the camp felt a lot lighter. We stoked the fire way up and never felt anything out of the ordinary after that, so we did sleep in the tent and nothing else happened that night.

I would like to go back out and just camp and act goofy and allow them to come back to us (if it was in fact them) so that we can make a trusting relationship happen. This is what I did at the Washington lake. Of course, I never felt this type of terror there and I camped alone there. I'm not sure I want to camp alone up on the lake above the Clackamas River right now.

So many fantastic things have happened at that Washington lake. I still try to go there as often as I can, but it's not often enough for me. I haven't been back since June (2010) as a matter of fact. But once I get my truck running again, I plan on taking a week or so to myself and going back up for awhile. This is the place where I cast the most perfect size seventeen and a half footprint in the mud underneath a water runoff pipe next to the road. It had a perfect midtarsal break and everything.

I would like to find another clan closer to home. Just keep going out like I did before, the more they see me there the more they will trust me. This is what I did at the Washington lake and the female and the little boy trusted me, but I think perhaps her very large, very red mate did not like it one bit. Just a feeling that kept coming to me out there, but I trust my feelings a great deal.

We could make it to the Oregon site before the snow flies, but that Washington site is three hours from me now and without a truck I'm grounded from long drives.

In response to the situation being experienced in the Clackamas River country, Kathi and I arranged a date to meet and to further investigate the anomalies being found in that region. I related to her: "I'm no great shakes at this, I just do like you do. I set up and let them come to me when they're ready... I read, I write and I listen with my entire being... I guess I'm pretty dull to all but them but they seem to like me..."

When we made our way into the mountains we found the conditions I have alluded to prior. There was trash in great quantities on virtually every road and at every wide spot on those roads. It was not the detritus of industry. It was not things left over from the loggers who worked in this area. There were car tires and batteries in the ditches and in the waterways. There was so much rotting paper and piles of plastic that I was literally appalled by the conditions here. Whenever we stopped to gather wood for our evening fires, we also cleared as much of the toxic items as we could carry. We had several car batteries and oil containers that people had seen fit to dump hither and yon. We couldn't carry the tires and but few of the plastics if they did not contain malevolent chemicals. In the course of a day, we literally filled Cristy's truck with this garbage.

We spent a most pleasant day in God's wonderful creation enjoying all it had to offer when we could ignore the people shooting nearly incessantly at, hopefully, some of the gravel pits in that area.

As one can imagine, the sound of a rifle carries a long, long way in this still, quiet biome and the reports were most obnoxious.

As evening gathered we enjoyed a particularly good dinner prepared for the occasion and passed the twilight in pleasant conversation among good friends. As dark descended, we made ourselves ready for the night… I in my small tent and the ladies in their tent. This night was cold. I had prepared for a cool fall night, but as night deepened, the temperature dropped significantly. By midnight I could no longer tolerate the situation inside my tent and retreated to my car. I warmed up enough to return to sleep and was resting comfortably when it occurred…

Kathi Blount

Suddenly I was awake. There was a pressure on my chest pressing me down as if a weight had been placed on it. I was struggling to breathe and I could not lift my arms. I was near panic when I realized what was happening. "Wait," I stated aloud to an empty car interior, "stop this and talk to me. We are not here to harm you or harm your home. We are here to learn from you if you would allow it."

A picture of the trash we had been seeing all day came into my head as a voice rang in my very mind… "Have humans no respect for their home at all?"

"I understand," I replied. "If you will look into our truck you will see that we have been picking up what of this trash that we could. We are only one small vehicle so cannot do much, but we do what we can. I agree, it is a terrible situation."

He answered, "We are not of here. We live higher on the mountain but have come here because the swimmers are much in the river. We will be returning home when they are gone. Do humans not know respect?"

I must admit that is a good question. The weight was lifted and my breathing returned to normal and I slept. Although the cold wakened me a few more times, I pretty much rested comfortably until the sun was up.

I did not mention what had transpired in the night until Kathi brought it up herself. When we compared notes, it turned out that we had both had the same experience and, while we could not prove this, probably from the same individual.

I returned home that day. I did not stay another night but Kathi and Cristy did. I would assume that the night went well because they were both in attendance at last May's Conference in Richland, Washington.

Redstripe Mindspeak

By
Thom Cantrall

Redstripe

"This place where you are going, is there a stack of wood alongside the road there next to the river?"

These were the words of my mentor as I we were discussing my pending trip to the upper Touchet (pronounced two-she) River in southeastern Washington state on 25 September, 2010.

"I see a large male with a red stripe on his chest looking

across a pile of logs at you. My teacher has said he is there to meet you."

Thus began the journey I continue today. I had responded that I knew not what lay on that road as I had never traveled it before. I had been on the ridge top on both sides of this canyon, but I had never ventured into the canyon itself. The land there belonged to the Indians. It had been deeded to them by the Boise Cascade Company and the Umatilla Reservation had created a public access hunting area that had only recently been opened for access by the general population.

I made my plans then to travel up that road as far as I could in my car or until I found this deck of logs my mentor was describing. As it turned out, it was not difficult at all to locate it. It was merely about a half truckload of logs that had been cleared from the road right of way when the improvements were done that allowed access. They had been conveniently stacked to be transformed into firewood by some tribal member so needing it. When I located this feature, I stopped and made that my base camp for the experience that was to come… an experience that was to change my life.

Pile of wood between road and river

Chapter 10

That "Still, Small Voice" Within Us

By Thom Cantrall

The late-day sun was peeking through the shimmering pines at a low angle, bathing the glade beside the small but energetic river in a strange, yellowish glow. The insects that had been so insistent but moments ago were suddenly still. In that instant a living, vibrant and noisy dell had transformed into the epitome of silence where the only sound to be heard was produced by the water of the stream tumbling and splashing its way toward the distant Pacific Ocean. Here, in the valley of the Touchet, time stood immobile while one could have counted to ten in a slow, measured cadence.

Suddenly, a whistle was heard. It was a whistle that one might have easily mistaken for some unknown bird inhabiting the brush and trees beside the stream. To the uninitiated, it may have been so... to me, who had heard it so many times before, it was a whistle no bird ever known to the Blue Mountains of Washington and Oregon had ever uttered. My eyes jerked around like a laser sight locking in on a painted target. I knew it was no bird, and, moreover I knew exactly what it was. As my eyes snapped to it, the scene came into focus. I saw him as he lowered his more than nine feet tall body out of sight behind a screen of brush. He was magnificent in his presence. I was seeing him, the object of my quest, at a range of less than thirty feet and he was huge. He was nearly all black with a massive, crested head that sat on broad shoulders with seemingly no neck although the turning of his head belied that. His upper body was reminiscent of an Arnold or a Sylvester in their prime only much more so. He was covered head to toe in long, very dark hair. Only his eyes, his nose and his hands were bare of this pelage. His eyes were so liquid brown; I felt I could have swum in them. In looking to him, an anomaly presented itself in that he had a darkish red stripe that ran from the point of his shoulder down across his chest and ended just above his hip on the opposite side. It was a mark of his status within his group, this special symbol… it set him aside as a teacher… my teacher, I was to come to learn.

"Where I Saw Redstripe"

It was obvious he was in the process of ducking out of site and I did not want him to leave. I had come to this remote place to find him. I had been told exactly where he would be and how I would know him and that had become true.

"Don't go," I uttered aloud. "Please stay and talk with me, if you could."

That was when it happened. Into my mind came a clear and distinct voice, "I cannot stay," the voice stated. "There are too many people here today and I cannot be found out."

There were other words spoken into my mind that day, but the content was not important to this narrative, only to me. What was important was the fact that a very large individual among a race of very large beings had spoken directly to me for the first time in my more than fifty years of involvement with Sasquatch. Not only had he spoken to me, but he had done so in a manner that I had experienced only rarely in my life. He had spoken directly into my mind! And, I understood it perfectly. I have heard these beings "speak" on several occasions before, but always one to another and it was but gibberish to me. Some seemed to be on the edge of understanding. It was like they had been speaking in an unknown but nearly recognizable language, but this experience transcended that. This was mindspeak in its purest form.

All my life I had been told of that "still, small voice" that speaks to us to warn us or to deliver a message to us and I had experienced it on occasion.

Grand Mound - pushed up by the snout of the last glacier

It was winter in western Washington and it was raining as is its wont. I had been to the small town of Tenino to visit a friend who lived there. I'd had dinner with him and his family and was driving back to my home in nearby Chehalis. It was not late, but it was well past dark at a time when that dark descended prior to five pm and the rain and darkened pavement made diligence a must on this night. The road was flat and straight for several miles across the dimpled mounds of the Mima Prairie. This was the spot where the glaciers had ended their southward march some fourteen thousand years ago and the melting of the ice had left a vast area of drumlins on the flat wasteland. This is the area of Grand Mound, the ridge of glacial debris pushed up by the snout of the glacier at the extreme limit in its southern advance.

As one nears the junction of the Interstate Highway and this back road to Tenino at exit eighty eight off Interstate Highway 5, the

The Mima Mounds - Drumlins left 12,000 years ago by the glaciers that ended here

straight stretch ends with a curve to the right to align the roadway as it crosses over the Interstate. The curve is not severe and can easily be negotiated, even on a rainy night, at the fifty-five mile per hour speed limit on the road.

As I was nearing this curve, I saw a set of lights coming up behind me at an extreme rate of speed. I was traveling at just below the speed limit, being in no particular hurry to return to an empty apartment after having enjoyed a splendid repast. As the pickup sped past me at an estimated speed in excess of ninety miles per hour, I edged as far to the right as I could go and slowed a bit to allow him all the room he needed. It was a standard cab pickup loaded with high school kids returning to their home in Rochester after a basketball game at rival Tenino High School. I merely shook my head at them and wished them luck as they would no longer be my problem, or so I thought at the time.

As the taillights disappeared around the curve ahead I eased back into the center of the lane and resumed my traveling speed. In doing so, I contemplated fools, teenagers and drunken sailors and how they survived in the world. Complacency had returned as I neared the curve when that still, small voice sounded in my ear, "Slow down, Thom, they are not going to make their turn."

Immediately, I took my foot off the accelerator of my truck and prepared to brake quickly as I coasted slowly around the curve leading to the Interstate. I was in deep fear for what I would find. It could have been mayhem if the truck had left the roadway and rolled. There could have been young people dead and dying and I did not anticipate this with any kind of pleasure, for sure. What I found was not what I feared. The marks on the road surface were plain in that the driver had lost control of the truck and had spun around a full three hundred and sixty degrees at least twice before departing the roadway backwards into the median area. Against all odds, the truck was upright and when I stopped to check on the occupants, they were scared witless, and their voices showed extreme excitement but they otherwise appeared well to me. I asked if there were any injuries, bumped heads or banged shins and was answered in the negative by both of the two young men and the two young ladies, while not so boisterous, indicated they were unharmed as well.

This, of course, was before the age of cellular phones so, there was no way to call for help on the scene, so, after having convinced myself that they were physically alright and that there was no sign of alcohol, I told them to stay with their truck and I would stop at the first business open and call in for help.

I left them to their own devices and traveled to a small grocery and made that call, describing exactly the situation and the location. I sat, then, to contemplate what had transpired. That voice echoed in my mind even yet. It had come to me with such clarity and brilliance that I had no doubt as to its veracity. I obeyed immediately. It had spoken to me as one person speaks to another except it was not audible to anyone but me; it was simply heard in my mind. I loved the fact that this was so and I marveled in the beauty of it.

The thought of this voice never left me, and I heard it from time to time in my life, always as a warning of something such as what was related, or as a comforter in times of great tribulation in my own life. It came to speak to me many times and I heeded its message diligently for I knew its source.

In late 2002, I was working at a sporting goods store in Kennewick. The job was fantastic in that I got to show and shoot bows and talk hunting all day long and they paid me to do it! The pay was not generous, but it was enough to live on and I was happy there. I had an apartment of my own that I had someone come in and clean on a regular basis. I had my little car and I was quite happy with my situation.

It was a Saturday and I was tired. I had worked a long day and had been busy all day equipping and tuning bows for customers. When I was finally able to leave at closing, I was looking forward to getting home, having a bite to eat and sitting in front of my computer to talk to some friends. As I was on my way home, I crossed the main north-south highway that ran through our town and that voice returned once more.

"Go have dinner," it said to me.

"I'm tired," I replied. "I just want to go home and rest up before work again tomorrow."

"Go have dinner," it repeated.

Well, I had long since learned that argument with this voice is futile, so I turned around and headed back into town. "Having dinner" at that time usually meant one thing... having Chinese. It was cheap and the local restaurant was very good. Therefore, it was my destination of choice on short notice.

As I took the seat assigned to me, I noticed a new waitress was working tonight. With a small, self satisfied smile, I wondered if I would meet her tonight. I need not have worried for she came directly to me. We talked and I ordered... she delivered my dinner and we talked some more. I learned that she was new to our area and was homeless and living at the local "mission" which was across the Columbia River from the restaurant. There is no public transportation in our area past 6:30 pm, so, while she had no trouble getting to work, she often had to walk the four miles or so to where she was staying. I had no idea of doing so, but I found myself offering to share my apartment with her. We got together a few times before she allowed that I had no ulterior motive in this arrangement. Eventually, she accepted the offer and we moved her meager belongings into my apartment. That was nine years ago and today she is still as close to me as my own daughter. Although I now live in a retirement community and she is in her own home, we visit often and have dinner together regularly... all because I listened when that still, small voice told me to go have dinner.

Where I Saw Redstripe

The voice I heard while talking to my red striped friend was of the same type of still, small voice as I heard in those other two instances, but the voice itself does not sound the same. We talk regularly now, Redstripe and me. He has no problem making me understand exactly what he is trying to convey to me. I love that he can do this and I look forward to the next time immediately after we finish a conversation.

How many others know this voice? I don't know. I would hope that everyone has this experience, but I don't know that for certain. Redstripe tells me that we humans once had the ability to speak in this manner ourselves, but we have lost it over the twenty five or thirty millennia of our span on earth. Perhaps it would be worthwhile to work to reestablish this marvelous form of communication, though, I must admit, I have no idea at this point how to start. Therefore, for the nonce, I will be content to receive and just listen to this voice and continue to hope that the day will come when I can "send" as well.

Vision

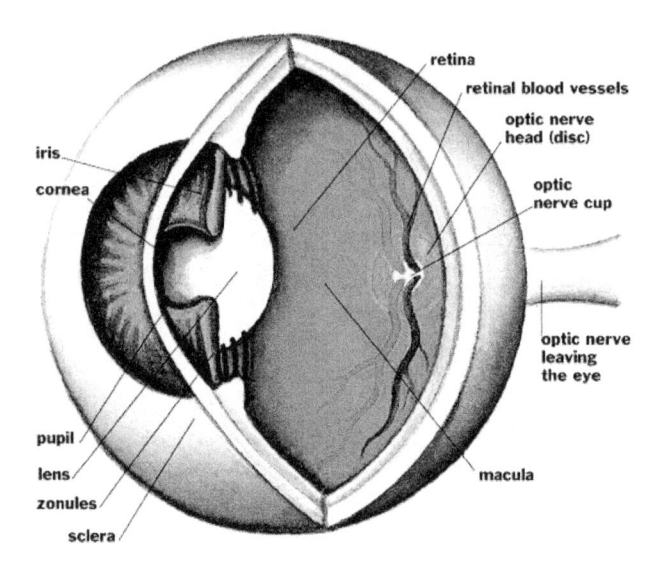

Eye diagram with labels: retina, retinal blood vessels, optic nerve head (disc), optic nerve cup, optic nerve leaving the eye, iris, cornea, pupil, lens, zonules, sclera, macula

    Once, many years past, I decided it was time to pass my "Advanced Hunter's Education Course" as offered by the state of Washington. I sent for the course materials and was most surprised to find, as part of the course, a study conducted by the University of California at Santa Barbara on vision… specifically, on how animals see as compared to us. This study was amazing in its presentation and I was genuinely impressed with the depth and dynamics of it. Probably the most elemental aspect of it though, was learning the various aspects of light and how different eyes see it differently. I was interested to learn, for example, that ungulates do not see into the red spectrum at all. Very simply, this means that these animals do not even see the hunter orange now required in most states for hunters in the field.

Equally interesting to me was the fact that they saw well into the Ultraviolet range. Of course, this would be a simple necessity of life for them as so much of their life is nocturnal. When one feeds at night, the time of maximum ultraviolet light, it is essential for survival that this be so.

Further study indicated that there are many creatures that see well into the Infrared range of the spectrum. It was therefore but a short step to see, knowing of his antics with IR cameras, that our large sasquatch friends could see well into this range as well. Since they often fed and operated nocturnally, it was a small leap to conclude that they must have well developed UV sensors as well.

This essay is simply a treatise I authored on how animals see… and, consequently, how we see. I present it here as a primer on the subject.

Chapter 11

Vision

By

Thom Cantrall

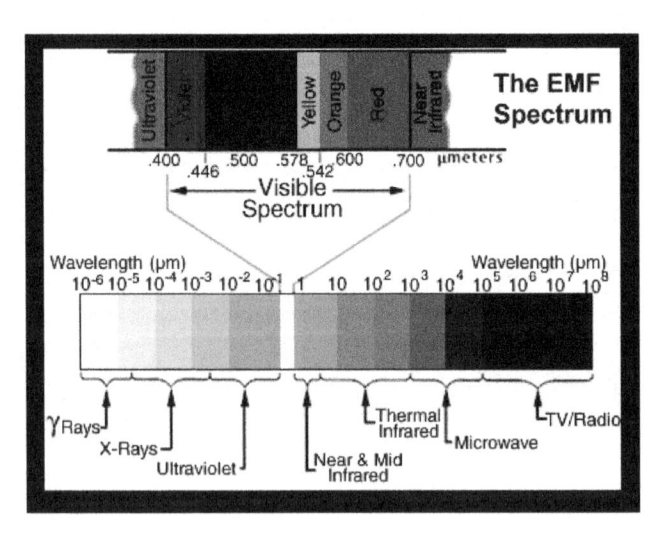

What is Vision?

Light is made up of energy waves of varying length. The shorter the wavelength is, the higher the frequency of the energy wave. When these energy waves strike an object, certain of them are reflected much the same as a ball striking a wall is reflected. The energy reflected then passes through the lens of the eye. The bulk of that light lands on the rear of the eye which, in most animals, humans and other primates included, is populated with tiny receptors called cones. There are three different types of cones found in the primate eye. The three cone types provide a highly refined form of vision. The remainder of the light falls around the periphery of the back of the eye where there are found both cones and rods. The rods are designed to receive the shorter wavelength, higher frequency energy levels. These receptors are designed to convert this light energy into an electrical impulse which is transmitted via the optic nerve to the brain where it is interpreted as a particular color of light. Many of these mosaics of light pulses are combined to create an image. In humans, the tri-chromatic vision yields a precise, well defined picture composed of many subtle shades of color and texture.

As one might suspect, the functions of the rods and cones are separate and distinct. The more prevalent cones are the only structures found in the center of the eye where most of the light strikes are designed to receive the longer wavelength, lower frequency light. They operate most efficiently in the range from a wavelength of around five hundred and fifty nanometer (nm) to the limits of our visibility around seven hundred nm. This encompasses light from the greens through the yellows and into the reds and is limited by our vision to just short of the infrared. In sasquatch, the limit is not as abrupt but allows them to see just slightly longer wavelengths, perhaps as high as eight hundred nm. This gives them vision into the infrared range which we, as humans lack.

Rods, on the other hand, are designed to react to light in the shorter wavelengths. They work, typically, below the five hundred twenty nm length, translating the lower end of the spectrum, the greens, blues and violets to the brain. The shortest wave length transmitted by these rods is around four hundred nm in our eye. In the human eye, the rods are fewer and are spaced more around the periphery of the receptor area. Also, in the human eye, the lens through which the light is channeled to the receptors contains a

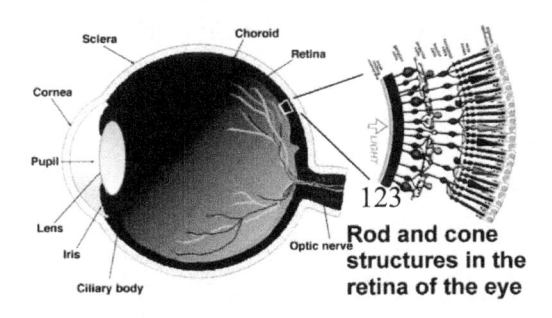

Rod and cone structures in the retina of the eye

heavy yellow filter whose job it is to remove light in the ultraviolet range before it enters the eye. This lens is variable in size as well, opening wide in low light and constricting as the brightness increases.

When light strikes an object, certain wavelengths are reflected, striking the lenses of our eyes. This lens then filters the UV and allows the rest to pass to the retina where the receptors are located. Most of the light, perhaps as much as eighty percent of it strikes the cone rich center area and the remaining twenty percent of the light strikes the periphery where the bulk of the rods are found. The light energy is converted to electrical pulses which travel to the brain where we now "see" the colors as determined by the wavelength of the light reflected. In other words, if light of five hundred and eighty nm wavelength is reflected, when all is said and done, we recognize it as yellow.

Seldom, however, is a single wavelength reflected. In fact, we can use Red, Green and Blue, the primary colors, to produce any color desired in our spectrum.

How Animals See:

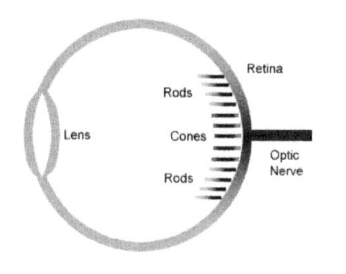

In Ungulates, deer, elk, cattle, etc, the basic structure of the eye is the same as in humans, but the arrangement is slightly different. The rod to cone ratio is much higher. Many more rods are found throughout the eye and especially in that all-important center area where the bulk of incoming light is received. Also, in these eyes, the yellow UV filter found in human eyes is nonexistent. An ungulate's eye lens passes all light waves intact. With his increased sensitivity to reflected waves, an increased number of rods designed to receive shorter wavelengths and no filter, it becomes apparent why these animals respond differently to the same stimuli to which we humans respond. In sasquatch the eye structure, while closely resembling ours, also contains more of these rods structures similar to the eye of the ungulate. It is this increased level of rods that allows him to operate so effectively at night. Further, the yellow UV filter is greatly attenuated to facilitate the passage of UV light to the retina of the sasquatch eye. These modifications are necessary to allow them the necessary vision in predominately UV lighted situations such as are found at night and in areas of deep shade.

The fact of a physical larger eyeball also compounds the effect. Since the reception area is larger, there are more receptors and, as mentioned previously, there is a higher percentage of rods, there is a large increase in the number of UV receiving rods. Further, with a larger eye, the ungulate and sasquatch has a larger lens. This lens can open at least three times the diameter of the lens in the human eye, thereby admitting nine times (three squared) the amount of light the human eye admits.

In the sasquatch and the ungulate, we have creatures that have a physically larger eye with a pupil that is able to open a much higher percentage of the total eye to the available light which is unimpeded by a heavy yellow UV filter as is present in humans. Also, their eyes contain a physically greater number of light receptors, a higher percentage of which are rods, sensitive to the shorter wavelengths of light.

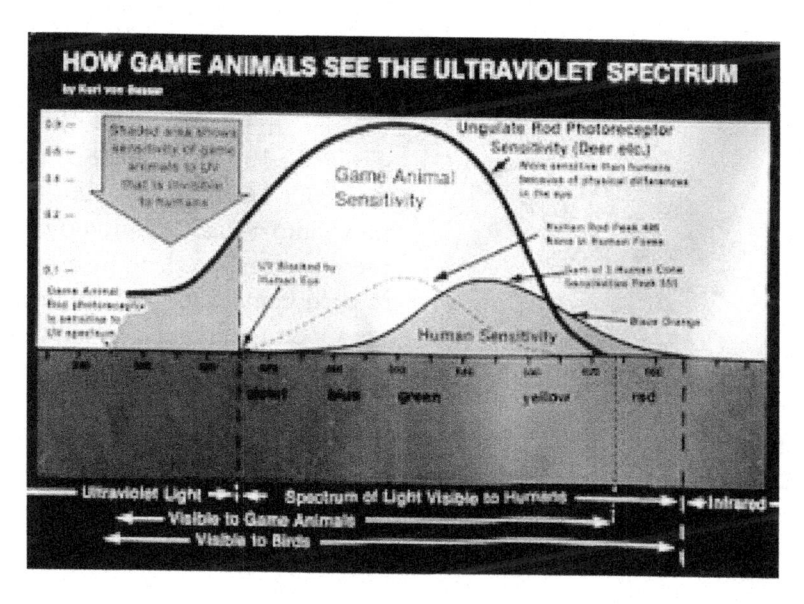

Graph of Vision Range in Animals

The figure above shows the energy levels visible to the human eye and to the ungulate's eye. For the Sasquatch, merely extend the longer wave length capability to the right in the chart to extend into the Infrared spectrum. Also, his vision into the shorter wavelengths would hold at a higher level, making him even more capable of seeing into the UV light range.

In addition, in ungulates, the cone construction is physically different, with one of the three types of cones not present. This has the effect of creating a dichromatic pattern versus the trichromatic pattern found in human eyes and in sasquatch eyes. Of all the mammals, only the apes and monkeys have this trichromatic arrangement. The net effect of dichromatic vision is the loss of reception of the red spectrum. This leaves these animals with, basically, two primary colors versus the three primaries we enjoy. The ungulate, therefore, has a vision range from four hundred nm to about six hundred and seventy nm. Due to physical differences in the eye, the ungulate's sensitivity to all light is increased by approximately nine times.

As can be imagined from seeing how the nature of our lighting changes to our perception, the energy of light varies in different conditions.

During bright daylight hours, the higher frequency, shorter wavelength UV light is diminished and the center of the spectrum predominates. As we move into a shaded condition these shorter wavelengths begin to predominate and the center of the spectrum diminishes while the lowest frequency, longest wavelengths, the reds and infrareds, are attenuated, but are prominent nonetheless. As we move into evening and direct sunlight is not available, only reflected light, the reds begin to disappear first with the yellows and greens following. The light available is high in the blues, violets and the, invisible to us, ultraviolets. Full night brings with it the loss entirely of the reds and yellows. Greens are a very low intensity level and the violets and ultraviolets predominate. It should be evident now why any creature that is active at night would require an above average ability to see in the UV range. At night, that is major wavelength of light available. Any creature not so equipped would be at a huge disadvantage and would not flourish in night time conditions.

Intermembral Index

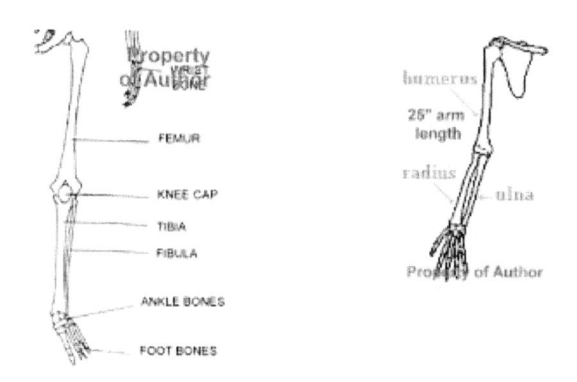

[Diagram of leg bones labeled: FEMUR, KNEE CAP, TIBIA, FIBULA, ANKLE BONES, FOOT BONES — marked "Property of Author"]

[Diagram of arm bones labeled: humerus, 25" arm length, radius, ulna — marked "Property of Author"]

Recently, the British Broadcasting company (BBC) funded a "documentary" that purported to disprove the Patterson-Gimlin Film. As with all such productions, not only did it not disprove the premise or substance of the film, but, in fact, tended to give more evidence as to the veracity of that special piece of film. In this attempt, the premise was that, with modern materials and technology, a suit could be constructed and worn is such convincing fashion as to "prove" the creature in Roger and Bob's film was pure fantasy. Pictured is a frame from that attempt showing a rather slender individual in a high quality suit making the famous turn found in frame 352 of the PGF. This is compared with the actual frame 352 image.

Immediately, two things jump out at us from the BBC film. First, the arms are much too short in relation to the body in the BBC suit. Second, the being in the BBC suit merely turns his head to look back illustrating that he has a very substantial and utile neck. A glance at the being in the PGF shows a much longer arm in relation to body length and a very short neck that forces her to open her shoulders in order to effect a look to the rear.

About two years ago I came across a factor that is profound. In fact, if understood properly by an intelligent person, this concept is legitimate scientific proof of the existence of the creature we call sasquatch.

Very simply put, the Intermembral Index is the ratio of the length of the arm to the length of the leg. The Index appears to be constant, within the standard bell curve in species.

In my own case, my arm is twenty-five inches in length when measured from shoulder joint to wrist. My leg from hip to ankle is 34.5". This gives me an IM of approximately 72.4... solidly in the human range. To alter that would require surgery I'm not prepared to undergo!

Study this concept and learn this chapter... it is one of the most important and compelling factors in understanding the sasquatch people.

Chapter 12

Intermembral Index

Dr. Jeff Meldrum of Idaho State University, Pocatello, Idaho, describes the Intermembral Index as the ratio of the arm as measured from the shoulder to the wrist, to the leg, measured from the hip to the ankle times one hundred. The one hundred factor is simply to clear the decimal from the result. That computation changes the result from a percentage to an index. Mathematically, that is Arm Length divided by Leg Length times one hundred or: **AL / LL X 100 = IM**

In primate species, all members have a distinct and specific IM. They break down as follows:

In a human, the IM is **72**

In a chimpanzee, the IM is **108**

In a gorilla, the IM is **122**

In a sasquatch, the IM is **84**

As can be readily seen here, the IM of human and sasquatch are remarkably similar but certainly not identical. It is this difference that can be used to make qualitative evaluations on reported sasquatch photographs and videos.

If we return to frame 72 of the Patterson Gimlin Film again, the arm length and the leg length are readily apparent and can be easily measured. Also, the arm and the leg are at the same distance from the lens in the same plane so no distortion to foreshortening is introduced into the exercise. The previously discussed calculation using the formula provided by Dr. Meldrum yields an ***IM of this figure to be 84***. That measurement and calculation places it firmly into the range expected from its species.

At this point the ardent skeptic, the sort that this essay is designed for would state something like, "Oh they just used an arm extender to make it correct..." Let us examine this more closely.

I measured my arm as best as I could in doing it alone and came up with the following data: Arm Length = 25"... Leg Length = 34.5"... therefore, my IM is ***25/34.5X100=72.4***... While this places me most firmly as human in structure, it creates a problem for he who would use an arm extender to achieve an IM of 84 as evidenced by the figure in the film. Solving the equation to yield a known IM is as follows:

AL / 34.5 X 100 = 84 or, <u>AL = 84 X 34.5 = 29</u>

This means, simply stated that my arm length of twenty five inches would have to be extended by four inches to achieve the desired result. In viewing the figure here, where would

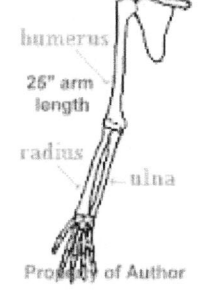

131

that four inches be added? If it were done as in the figure above which shows the arm extender, that would make the lower arm well out of proportion to the upper arm. In Frame 72, it is very evident that the elbow is exactly where it should be placed in the arm. As has been stated prior: "The elbow cannot be relocated." It simply is where it is.

A common ploy to trick the eye is to use oversized hands in the form of huge gloves to convey the idea of a longer arm. There was a recent video released by a fellow attempting to "disprove" the film… used this ploy, but the result was so emphatically terrible, it is beyond reason to ever believe it could be even remotely possible. Look at the hands in the film and it is readily apparent that they are in proportion to the rest of the body, not some outsized grotesqueness perpetrated by one who would have us believe his tripe.

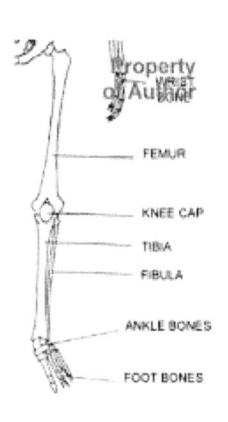

There is one other method for achieving the desired ratio. Perhaps one could reduce my leg length to the requisite twenty nine inches by somehow removing five and a quarter inches from its present length. Somehow, I think that might be even more objectionable to me than the process of adding the needed length to my arms. In short, there is really no feasible way short of surgery to alter this Intermembral Index formula or measuring criteria.

Humanity of Sasquatch

Much has been said about what and who sasquatch is. Some have maintained he's a relic hominid left over from the ice age that has somehow eluded detection for twelve thousand or so years while supporting an intellect about equal to that of a chimpanzee. Others maintain that he is merely an undiscovered species of great ape… that he has no more intelligence than, say, an orangutan or a gorilla.

In truth, I have discovered through observation and interaction as well as the gathering of evidence that they are much more intelligent than these people, university professors among them, know. They are not only sentient... heavens, a sparrow is sentient... but they are sapient beings. They are entirely capable of making judgments and determining a course of action from the situation as it exists.

In this essay, I will examine some salient points that illustrate this premise. We shall examine language, art and culture as they apply to the case at hand.

Above is the logo for the International Society for Primal People... Sasquatch... which I started to educate the public about this people's right to exist and prosper. I have received a lot of flack from certain areas and people over my views and that is alright. As long as a person remains polite and in respect, I will do the same and we can exchange ideas and positions without rancor. After all, we learn from people who have different understandings than our own. If we converse with only those who think and act as we do, how could we possibly learn new ideas? We learn when an idea is brought before us that we have not considered and we then analyze and test.

Chapter 13

The Humanity of Sasquatch

By
Thom Cantrall

It was mid June of 2010 and I was sitting in a lecture hall on the campus of Lane Community College in Eugene, Oregon. The occasion was the Oregon Sasquatch Symposium and the speaker was R. Scott Nelson, Cryptolinguist with over thirty years experience, including more than twenty years with the United States Navy. Scott was prefacing his presentation with some background remarks when I heard him say… "My son wanted to write a paper for his class in school on some phase of Bigfoot. I suggested we get onto the internet and search a bit to see what we could find. A website came up that featured recordings done that were purportedly the conversations of bigfoot creatures done in the mountains of California from a hunting camp. I listened to the sounds and I told my son, 'THAT IS LANGUAGE. They are speaking a language there'!"

As a crypto linguist in the United States Navy, Scott listened to radio transmissions from various people around the world, most of whom were our potential enemies. While Scott speaks three languages, including Farsi (Iranian) and Spanish, he does not speak Chinese nor Korean. If a transmission comes in that is in a language Scott does not speak, he must be able to determine if it is language or if it is gibberish sent to confuse and confound any who might be listening in. If he were to hear a transmission in, say, Korean, which he does not speak, he would be required to determine, from listening to the structure, etc. if it is language. If he so determined, it would then be forwarded to those who did speak that language to do the things they did with it. This ability placed Scott in a perfect position to understand that what he was hearing on the Berry-Morehead tapes was, indeed, language.

USS James Madison

Over the ensuing years, Scott has continued his investigations into the audio files compiled by Ron Morehead and Alan Berry in the Sierra Nevada near Lake Tahoe. He has developed a phonetic alphabet for their sounds by dissecting the morphanes (basic building blocks of language) and analyzing their structure in the overall scheme of their usage. It should be noted here that Scott stated that before he could begin to make any headway in this endeavor, he had to slow the tapes to less than half their original speed.

Ron Morehead

Language, especially in the form of articulated speech is a purely human trait. Within our throat, just forward of our larynx is a rather unique bone known as the Hyoid Bone. While it is found in several animals, including the great apes, its shape is unique in humans. This bone is also unique in the body in that it has no skeletal connection to any other bone. It is held in place strictly by muscles and cartilage. (This accounts for it being omitted from the old song about "The Knee Bone is Connected to the Thigh Bone", etc) the Hyoid bone, according to biologist John Morely is what allows our tongue to articulate speech. No other commonly known primate has this ability. Yes, chimpanzees and gorillas can make noises. They can even be taught some communication skills, but they are simply not capable of articulated speech. Articulated speech is the domain of the HUMAN animal alone.

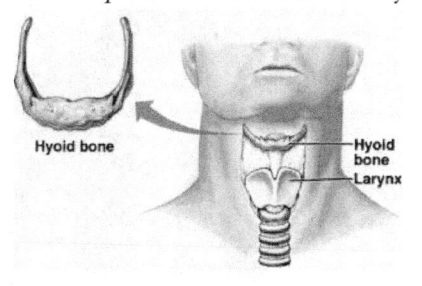

----- * -----

April 2011 found a small group of people gathered at an eastern Oklahoma homestead where a drum making workshop was being presented. In the evening, the hostess had determined to take a walk into a wooded area of her property to just enjoy the evening. Stewart, a federal law enforcement agent present for the instruction, asked if he could accompany her. A short hike found them out of sight of the campfire, but still close enough to hear the muted shufflings of camp life. Although the sun had long since set, there was ample light from a large moon to see well enough that a flashlight was not needed. I don't know exactly how the scenario unfolded in the beginning, but it found these two people in a position where they could see a group of sasquatch people quite clearly in the moonlight.

Stewart was asked, telepathically, not audibly, by the sasquatch people to move away from his hostess and stand alone. He complied and was asked almost immediately, "Why do you have a gun?" It is important to understand at this moment that the gun was concealed on his person, not in an open holster... It was not visible now, nor had it been removed from that concealment in the entire time he had been in attendance at this workshop.

"I am a Federal Agent," Stewart answered. "I am required by my position to always be armed. I carry this gun for this reason and no other."

What followed was recorded and is available for hearing at any time. The leader of the group of sasquatch people, nicknamed Owlcaller, began to audibly imitate the barred owls that inhabit this biome in abundance. While the pair of humans watched and listened, Owlcaller performed a cacophony of audio dynamics that tended to boggle the mind. The show was sufficiently loud to draw the attention of others at the gathering and two came close to join in the festivities. One, a deputy sheriff from a northeast Texas county answered the sasquatch with his own call and was rewarded with clearly heard audio responses. The second person moved to a position where she could watch a young child and an older sasquatch as this symphony played out in the beautiful Oklahoma spring night.

----- * -----

In the spring of 2012 there was a conference on all things Sasquatch. This conference, The Pacific Northwest Conference on Primal People (Sasquatch) was held in Richland, Washington the first weekend in May. It featured speakers from across the country and, more importantly, across the spectrum of Sasquatch Investigation. It was during the keynote speech at this conference that a new concept began to formulate in my mind. Our Keynoter was a beautiful red-haired artist from the east who consented to enlighten us about art in general and especially how it pertains to this new people.

I began to wonder, during this discourse, what form the art of this people might take and decided to formulate a hypothesis that I might test if not empirically, at least emotionally. I thought, after all, isn't art itself far more emotionally based than empirically based? How would one even begin to place an empirical scale on visual art to begin with? Could we assign numerical values to art? I can imagine the squabble that would ensue if I attempted to rate, on a scale of one to ten, the "Mona Lisa" or "Blue Boy"... or, how about Michaelangelo's Sistine Chapel? I could not even begin to imagine how one would classify ANYTHING from Picasso.

Tot 'n Tater

Oh my gosh, there would be war in the blogs! There would probably be a new group on facebook titled something like "The Coalition for Reason in Digitizing Art". I can only image Art Evidence stealing ideas from Renoir Lindsay about the rating system and Gaugin Dyer would be perpetrating all sorts of hoaxes on the art community... in one he most assuredly would claim to have sliced a Dali to ribbons! I cannot imagine the depth of it all...

It was at this conference, however, that I met a man I'd been chatting with for a time from British Columbia, Canada. While I do not at this date remember exactly how it began, we had a conversation on their art. Over the years I have seem many structures and glyphs created by the sasquatch people and have always marveled at them, but it was not until I began this conversation with Brian that an idea began to formulate in my mind.

Even still, it was from May until July or even early August before the evidence began to amass in quantity. Brian began sending me photographs of the glyph formations he was seeing and I began to catalog them... In time, I began to assemble photographs from other areas as well to the point where I now have well over three thousand such in my possession.

Asterisk

For several months I had an idea to prepare a Power Point presentation on this art work and began to construct in my mind the elements necessary to effect this endeavor. I was amazed at how difficult this assignment was becoming. The sheer volume was daunting and the responsibility was being driven home relentlessly. The task seemed to be expanding even as I waded through the surf of the subject. When I began to actually numerate possibilities, I found I had over a thousand photographs from British Columbia alone. Added to that were photos of glyphs and structures from almost every state in the union as well as my own pictures. By the time I was ready to select those I wanted for the PP Presentation, I had to choose from well over three thousand pictures.

If a journey of a thousand miles begins with but a single step, so it was here. My journey to this presentation began with a selection process made a bit easier by creating several categories and separating my albums into these several categories. Of course there were those that didn't fit into any real category but seemed to be crying for inclusion into the final product. When this first separation was complete, I had weaned the selection down to a mere nine hundred or so photos. Not a huge step from there to the final fifty or so, was it? ...Especially when one considers the fact that I had already eliminated all of those not worthy of inclusion. Now, it was just a matter of choosing the best of what remained.

Finally I declared the project complete and closed the book on it with just over three hundred photos included... and that very night I received photographs of some of the most artistic work I'd ever seen them do. Then, a few days later another person sent me an example of a beautiful structure. Just a week later, photos came in of a glyph being altered daily by a young adult male sasquatch person... So... I guess it's not complete. I am afraid it will be a work in progress forever but I can say the basis is complete and that allows me to draw some conclusions.

First, and foremost, the evidence is clear. The sasquatch people create and utilize art. I have, to date, identified and have had identified about seven signature glyphs. These are artistic designs that identify the creator. I have also identified or have had pointed out to me about twelve to fifteen other glyphs that I can interpret quite easily. These include the Friendship X, the Asterisk and others. Most are done in a matter of fact manner but some few are so beautifully done that they are almost purely art in their purpose.

Lest we think art has no part in the written word, I would refer to the manuscripts of medieval times and the fantastic artwork included in the lettering of many of those hand-scribed tomes. There is no doubt that many monks were very talented artists and they shared this talent in the manuscripts they produced.

Recently, on discovering the importance of these signature glyphs, I wondered what would happen it I were to develop my own signature glyph and share it. I then expanded that to certain of those with whom I work. The results have been nothing less than spectacular! In EVERY case, there has been a marked increase in correspondence and a similar increase in activity. Brian in Canada found out early on that they did not want his glyph to be left on trails or paths. They also did not like his initial attempts. An associate of Brian had his glyph altered and when they returned, written in sticks is a glyph that reads: "U R NEXT" ... I saw the spot this occurred... and I do believe the ONLY human who could have changed this would have been one who knew it was there. It would have taken prior knowledge as it was NOT in an area that would have been walked through at random. It was off the trail and in a very difficult area to access. Rather remarkable to say the least!

By far the most spectacular example to personal glyphs being left, however, comes to us from a team in far northern Canada. When they followed the advice to try leaving a personal glyph they were gifted with the most spectacular gifts hand woven from what appears to be some type of root runners. The three examples are nothing less than fantastic.

In the previous few pages we have discovered that the sasquatch people have a spoken, articulated language. Scott Nelson and Ron Morehead have provided us with unequivocal evidence and expert testimony to this fact. We have seen that it requires a Hyoid Bone in order to have this articulated speech. It is a given fact that none of the giant apes have the proper shape of Hyoid Bone to allow them to articulate speech. It would seem logical at this point to assume that since sasquatch people have been observed and recorded in this articulated speech, they would have a Hyoid Bone very similar to ours.

Clan Leader's Signature

In the 2011 incident in Oklahoma, this cultural display fostered a capability to discern danger that was yet hidden. They then conversed directly with the individual that presented a possible perceived threat and mitigated that threat to their satisfaction while displaying a behavior that kept the subjects amused, entertained and amazed. I know of no great ape or other non-human animal that is so capable. I do stand ready to entertain anyone who has an idea of such, however.

Work of Art

In my studies of the glyphs and structures these people make, I find pure art to be an integral part of it. I find creations that could, and probably should share space with the art of the day coming from the human realm. I know of NO other creature that creates pure art at random. Yes, the spider spins a beautiful web... but ALL MEMBERS of that species of Spider spin the identical web. Many birds sing a beautiful song, but ALL MEMBERS of that species sing the exact same song. Many birds build exotic and beautifully artistic nests, but ALL MEMBERS of that species builds the same nest. With my studies, my subjects build their art as they see fit. They create glyphs to display what they want to convey....

In conclusion, ONLY HUMANS have articulated, spoken language. ONLY HUMANS are capable of discerning hidden danger and neutralizing it thereby displaying a culture. ONLY HUMANS create art simply for art's sake.

This said, it is now abundantly clear that the sasquatch people cannot be anything but, essentially human in nature. We are Homo sapiens sapiens... Wise, Wise Man... They are Homo sapiens hirsutii... Hairy Wise Man... make no mistake about it, they are THAT close to us.

Coppei Homesteaders

A few years ago I came across a very well written, essay about the Coppei Creek area of eastern Washington. This is an area I know very well as it is the domain of a clan of sasquatch people I work with very closely. Coppei Creek is between the towns of Walla Walla and Waitsburg, the first incorporated town in eastern Washington, dating back to the middle 1800's.

I have a deep and abiding interest in the writings and opinions of those who lived here before me. What the pioneers found and coped with is always of profound interest to me and to this end I have read a large number of personal journals of those who were there and who did the things that made us what we are today. Here is one such personal statement as recorded by a former journalist in the eastern Washington area.

Vance Orchard (1917-2006)

I met Vance Orchard about twenty years ago when I read an article he had written on Paul Freeman that appeared in a Washington newspaper. I asked Paul about him and Paul brought us together. Later, after he had fully retired and was living in Spokane,

Washington, I saw this essay and again held intercourse with him. I found him an articulate and fascinating person. Vance is gone now, as is Paul… and, I suppose, it won't be that long until I go to meet them again. It will be a fascinating reunion, for sure…

This following is from Vance himself and explains a bit about the genesis of this essay:

"I have written about the people, places and things of SE Washington and NE Oregon for 46 years. For 38 years I did a column for the daily Union-Bulletin of Walla Walla and for the past seven years for the weekly Waitsburg, Wash., Times. The following story appeared in 1992 and was one of the most interesting! …and I have written many Bigfoot stories." – Vance Orchard

Paul Freeman with Hand Cast of a Bigfoot taken from Upper Dry Creek above Dixie, WA. April 21, 1994

Chapter 14

Coppei Homesteaders
An Account of Life Lived With the Sasquatch People

By Vance Orchard

Of course the call got my attention, especially when the woman ... in her 80s ... said the February Bigfoot track chase reminded her of the days on the upper Coppei when her family "lived right among them." Yes, and several other families who were homesteading in this section also experienced the family (or families) of Bigfoots who lived nearby, my caller informed me. My caller, by the way, remains anonymous in this relating. That's her choice, but the story she tells is one of the most interesting of the many Bigfoot tales I've heard from this part of the Blue Mountains. And, after all, the stories recounted about the subject are what make up a big piece of the interest this anomaly seems to hold.

Coppei Creek Homestead

My caller had opened the conversation by observing that a comment I'd made in a recent Bigfoot report, about Bigfoots possibly eating a deer killed by a cougar, would prove a point she remembered about her childhood days on the upper Coppei. There, at Coppei Falls, she said, Bigfoots reputedly would chase elk and run them over the brink at the falls. Then, at the base of the cliff, the critters could pack off the elk at their leisure! When I reminded her that early native American mankind had used the same tactic to kill off thousands of buffalo, she felt certain the old stories were true. And, she had a number of stories to tell, too, about this area in the foothills of the Blue Mountains near Walla Walla. The story of the Bigfoot encounter at Huntsville (Twenty-five miles east of Walla Walla) in 1900 has been the oldest recollection locally of the Bigfoots of the Blues. What my caller was talking about were the experiences of her family and other homesteading families of the 1920s. She said she was thirteen ("...that's seventy years ago") when her father recounted his own first experience with the Bigfoots which were to become so much a part of their life on the Coppei.

Coppei Falls

"My father told me one day he was looking across to an open hillside on which rested a huge white log.

Robinette Ridge from Coppei Creek

Then, he says he saw the 'biggest, strangest bear ever' walking out of the brush ... two others followed and then all went to the log and sat on it, then went to tearing it apart and had their lunch (of grubs, etc.?) then went off down the trail. My father made me promise not to tell my mother about this incident."

My caller was to have her own experience and encounter with a Bigfoot as did her mother and two sisters, I was to learn as the telephone conversation continued. What my caller saw that day possibly was a young Bigfoot. "What I saw was a sort of man-creature, about six feet tall, reaching up into a young tree, its fingers spread ... it had brown hair and its face was gray...it turned its head on its shoulders like an owl would to see me. I froze and it just disappeared into the brush." Her mother's encounter came one day when the family's herd of six milk cows was apparently "rustled" ... they were not in the usual place they were pastured in, my caller said. "We always figured the Bigfoot family had herded the cows off... mother followed the tracks and after a time caught up to our cows with a small Bigfoot with them. She figured the older Bigfoots had left the young one to look after the cows. Mother said she looked at the Bigfoot and it looked at her. She was scared but ran after the cows and she and the cows all ran toward the homestead. Mother said the Bigfoot was just a caveman, but he was dressed up in furs and really needed a bath!"

My caller's second encounter was not with the creature itself, but it must have been close by for the strong odor was noted, a factor in many sightings of Bigfoots. "My two sisters and I had ridden horseback into the woods one day ... we came across where limbs had been stripped from a young tree in the trail. The horses wouldn't step over them ... there was a rustling in the brush ... my horse reared up on its hind feet and wouldn't go over the pile of limbs. Then, I smelled a strong odor ... like a hundred pairs of socks from as many stinky feet ... not a 'dead' smell and not a 'skunk' smell ... a sweaty smell... real pungent, too."

My caller made an observation about the Bigfoot family and its possible relationship to the abandonment of homesteads on the Coppei. I have an opinion on that, but it would be interesting to talk with some of the descendants of those homesteaders. Members of the Bigfoot family often raided her family's garden, my caller told me. "Many people around there had experiences of following tracks made by a big barefooted man .... Sometimes the stride of tracks was so far apart we had to run and jump to get from one step to another ... it was a real big one, I guess." My caller also noted that "a real big one looked into our windows one night. Lots of people up there had this sort of experience and it was pretty unnerving to them ... I think it was a reason many left that area and gave up their homesteads ... they just moved out and never said why, although not all were doing too badly." My own idea of why some people gave up homesteads was simply because the land would not provide a living for the family. I know this was the experience of my mother's family in the Colville country at the turn of the century. When the homestead was "proved up" her father had to head to the nearest town and take a job in a sawmill to earn a living. The land wouldn't do it and nobody was buying the land, either. So, the family wound up just moving out and abandoning the plans for the homestead, my mother told me.

A couple more comments about the Coppei Bigfoots. According to my caller, the creatures "were a family ... no mistake about that and I think they even had names for each other. And, another thing, when a Bigfoot got angry, they'd throw rocks at whatever made it angry."

Lake Roosevelt Encounter

Kyle Gibson

    Summers are hot in eastern Washington. Most contractors work hard all summer while the weather allows as the winters can be as cold and rugged as the summers are not. For a contractor to be able to steal some time away from his business is a rare and treasured thing.

    Kyle Gibson is no exception to this rule. His landscaping business keeps him running from before daylight appears until long after it is a mere memory. Because of his hard work and diligence, he has been able to create a small window for summer recreation. To effect this, he has retained a camping spot on Lake Roosevelt, the lake behind the Grand Coulee Dam that runs all the way into Canada. Kyle and his wife, Onette have installed a camper on their campsite and maintain a boat there as well. By doing this in this manner, when the opportunity arises, the couple can leave town at a moment's notice and drive the two plus hours to the lake where they can relax before the pressures of running a business draw them back to town and their crew and customers. Kyle is a first rate operator and Onette runs his business ends and, using computer graphics software, prepares proposals for prospective clients. This past year, this team took top honors for their design at the Tri-Cities "Parade of Homes" competition.

Landscaped Yard

Kyle has long been a sasquatch experiencer of the first order. He has a great deal of experience with these enigmatic creatures all over eastern Washington. I have accompanied him on multiple outings and have found him to be knowledgeable, resourceful and, most importantly, a man of integrity.

What follows is Kyle and Onette's own story. It is told exactly as they related it to me. I have been to the area described and have pictures from there. It is a quite unique area, to be sure. I had my own encounter there just in June of 2013 so I know the sasquatch people are living there in those environs.

Chapter 15

Voices on the Lake

By

Kyle Gibson

As Told To

Thom Cantrall

August twenty-fifth dawned warm with a promise to get even warmer in this far, remote area of north-central Washington state. Taking advantage of a rare day free of the demands of a growing business, we were happily camped on our site near the tiny town of Lincoln, Washington on the banks of Lake Roosevelt.

Lake Roosevelt stretches some one hundred and fifty miles behind the Grand Coulee Dam all the way into Canada. In fact, eighty eight percent of the area drained by the lake is in Canada. The land around our campsite is semi-arid with an annual average of about ten inches of precipitation. Consequently, this area makes for a very special summertime recreation area. The waters are never more than about sixty degrees even in the heat of summer, so a quick swim is always a refreshing thing. The land surrounding the lake is rough, steep and rugged.

There are pines and firs growing in the shaded areas of the north facing slopes and in the deeper gorges where the glaciers left enough soil to support the growth of a tree, but, for the most part, rock predominates. As the illustration shows, massive rock scarps rise directly from the lake and form a formidable barrier to egress from the lake in most areas.

Much of this lake's shoreline further north from out site falls under the jurisdiction of the Federal Government but in our area, it is under local control. The east side of the lake is mostly under private ownership with large ranches reaching to the edge of the water in many cases. The west side is all part of the Colville Indian Reservation and us managed by the tribes thereon.

Wildlife abounds here with mule deer, whitetail deer and bighorn sheep predominating. Recently the moose have returned to the area with cougar and bear being in ample supply here. Myriad smaller species are found everywhere. Of course waterfowl abound here as do both bald and golden eagles as well as osprey and other raptors. Grouse and Mirriam turkeys are everywhere. The waters abound with fish of several species including both large and smallmouth bass as well as the kokanee, a landlocked sockeye salmon. There are trout resident here as well.

Bighorn Sheep

It is certainly not difficult to understand why people are drawn here but being as remote as it is from major population centers, and due to its vast size, traffic is amazingly light and other people are rarely seen, let alone allowed to disturb one's quiet relaxation.

This Saturday began as many others had. We had decided to take the boat for a run down the lake a ways to a private little place we liked on the western shore. Being on the western shore assured us of privacy from the landward side as it was land belonging to the

Colville Tribes and was off limits to general incursion by tourists. The shore was particularly rugged in this area and we were hoping to have the opportunity of locating some of the plentiful California Bighorn Sheep in the area. Enjoying a calm day, we fished in a desultory manner, catching several of the very delicious Kokanee Salmon that were plentiful here as well. As the sun drifted inexorably west in a sky of perfect blue, we were at peace with the world when I got the idea to play Ron Morehead's "Sierra Sounds" audio CDs with the idea of seeing if we could get a response from the craggy mountains that bordered the lake.

As we drifted along, catching a fish now and again and listening to the strange vocalizations being played at a pretty high volume so as to effect a broad coverage, we were suddenly startled to hear an answer come from far up the mountainside. At first, I thought it might have been an echo because it was so faint, but there! It came again.

It was two diligent people who then attempted to locate the individual responsible for those sounds coming back to us from the mountain. Using my binoculars, I spotted him and watched as he moved slowly down the rocky bluff. Slowly he came on to the sounds. I can only imagine what he was thinking or feeling as he heard the vocal utterances from his type coming from the lakeshore. Was he thinking this was an interloper, come unbidden to his land? Perhaps he held the hope of finding a new mate? We have no way of knowing the why of what he was doing, but that he was doing it was now beyond doubt.

Kyle and Onette

Anxiously, we watched as he climbed down the steep banks until he disappeared... we knew he was moving on pretty rapidly and we had timed his movements, so, we thought we would know about when he should reappear down the slope further. It did not happen. He had disappeared and we had lost our chance. It was highly disappointing and a relief all at the same time. I was anxious to see him again will Onette was less convinced. Finally, I determined that if he would not descend further, I would go to him. We could see very clearly all around where he would have to pass in order to egress the area we had last seen him and nothing had passed in our vision. I was sure he had to be there in that low area somewhere and I was just as determined to see him again as he obviously was determined to remain in hiding.

We moved the boat in the cove to about a hundred yards from the rocky shore and I dove in and started swimming to the beach. While I was swimming contentedly along, unbeknownst to me, the big fellow reappeared to Onette still moving down the rock bluff towards the water's edge. Although she tried her best to holler at me and let me know what was happening, I could not hear her while swimming and continued along at a good clip… down the hill the creature came and inexorably to shore I came…

As I reached shore, I stood up… and looked a very large, very black, very powerful being directly in the eye at a range so short I had no real wish to be here. My guest solved the problem as he looked at me by screaming a very loud, very blood-curdling scream and I was frozen to my spot. I could not have possibly moved a muscle. I don't know if he had immobilized me or if I was simply in a trance, but I know I could not have moved a muscle. I do believe if someone had shot me, I could not haven fallen down, I was that rigid. While I was in this state, my friend retreated back up the hill and out of my range of vision and I was released from my paralysis. I immediately jumped back into the water and swam my fastest back to the boat. Those reports from my wife that stated I was up and planing are an exaggeration, however! Fast I was... but not that fast.

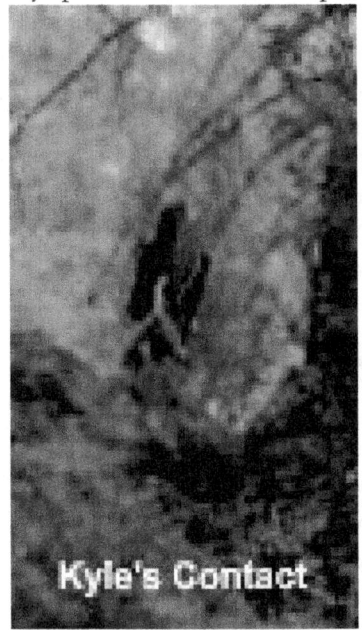

Kyle's Contact

Fortunately, Onette was able to get a picture of the fellow, and even though he was far away and the image had to be enlarged greatly, he still shows very well in it.

Hero Report

Every word presented in this volume to this point is absolutely true as reported. Of course, I was not present for Sargeant Neiss's encounter, Missy Hutzler's, or any of the others reported to me, so I cannot testify for aught but the integrity of the those who did so report. All encounters of mine are correct and true as reported.

... That is true of all those that have gone before... it is not true of the report to follow. Have you ever wondered what the sasquatch people think of us and our feeble attempts at investigating their activities and lifestyles? Well, what do you suppose would happen if they had an investigatory agency like some of those we have in our society? This is the story of what could happen in that case!

I cannot even give you an author here, or cite you a source. All attempts to identify a genesis for this have been in vain. This was sent to me intact by a dear friend of mine and I am reprinting it here simply as a change of pace. If this is your work and you object to me using it in this form or format, let me know and it will be removed post haste. Otherwise, please enjoy what I feel is one of the best written pieces of sasquatch humor I have ever read.
Thom

Chapter 16

H. E. R. O.

Human Evaluation and Research Organization

Interim report of Expedition 041206.

Author: Harold S. Harefoot

When the Sasquatch Intelligence Agency (SIA) learned that the 'Happy Wanderer Hiking Club' of (**deleted**) had planned a three-day outing to a primitive campground in area known to us as Honeybear Mountain, the decision was taken by the Investigation Committee to send a team to cover the event. The general location is in the southeastern United States (Exact location undisclosed due to the potential for ongoing investigation). Our team #3 was given the assignment.

Team Members:

Males:

Harold (me)

Herschel

Nigel

Morris

Females:

Cassandra

Elvira

All team members are experienced and highly trained researchers and investigators.

Equipment:

At times, we find it to our advantage to make our presence known to humans without showing ourselves. For this we rely on 'The Stench'. We obtain this from Simply Nauseous InFusions (**SNIF**), Ltd., who make the product available in handy aerosol cans and in a number of custom blends. Our choices for this exercise were:

Females: Cass and Ellie chose "FEMME" (damp forest, day-old garbage and honeysuckle)

Males:

Herschel, Nigel and I selected "MILD MALE" (wet dog and musk with a hint of rotten cabbage)

Morris, our extrovert, opted for **"BARF!"** (skunk, ammonia, rotten eggs, open sewer, rotten meat, rotten fish and gorgonzola)

All investigators also carried an aerosol can of **SNIF**'s tried and true "Narcissus Pheromone". One whiff of this odorless compound gives humans an uncontrollable feeling of 'being watched', causes the hair on the back of their necks to stand up and makes 'goose bumps' rise on their arms.

Objective:

As we have been observing humans for centuries, we believe we know most of what can be learned about them. However, we continue the program to keep pace with any improvements in their equipment and also to document any previously unobserved human characteristics. It should be noted that, although their stress levels are purposely elevated from time to time, particular care is taken on our part that no injury comes to any human during these exercises.

All investigators were cautioned that they were not, under any circumstance, to allow themselves to be photographed by humans as anything more than amorphous blobs in the wilderness. It is realized that some of our kind think it would be to our advantage to allow ourselves to appear crisply and clearly in photographs. However, we are staying with our principles. This policy will remain in effect until humans relent and guarantee us fair and reasonable compensation for our time and effort in posing for them.

Preparation:

It was known that the humans would begin to arrive mid-morning on Friday. In order to re-familiarize ourselves with the terrain and trail/campground layout and condition, we arrived Wednesday afternoon. The area was in excellent condition so we were able to complete our preparations by Thursday afternoon leaving sufficient time for an evening of fellowship. Beginning with a rousing sing-along, we then spent the rest of the evening telling human stories and jokes.

"Why did the human dash across both lanes of the busy Interstate?"

"Because he saw Morris on his side of the road!"

(Morris laughed so hard at this that he choked and a piece of the rattlesnake he was eating came right out of his nose!)

Before dawn the next morning (Friday), Morris was dispatched to the paved road to obtain some specimens of suitably-aged roadkill for use in an experiment we wished to perform. He found several excellent samples. In spite of eating most of them on his way back to our position, a sufficient amount remained for our purposes.

At 0930 we gathered to coordinate our first evolution.

The Arrival

At 1030 A.M. the club members began arriving. We observed carefully in order to evaluate their camp making and general forest skills.

Observed:

Fifty percent displayed *average* abilities

Fifteen percent received *above average* marks.

Ten percent were deemed to be *superior*.

Twenty-five percent should not have been allowed in the forest unaccompanied.

Conclusion(s):

These data conform closely to previous observations. Nothing new was learned from this exercise

The next exercise was to evaluate the human powers of observation and determine gastronomical preferences. After the campers had finished their evening meal and were gathered for informal social activities, Herschel, quietly and unobserved, made his way to the edge of the camp clearing. He tossed a portion of Morris's properly-aged roadkill into the clearing, made a couple of bird calls and silently withdrew. Immediately, three campers (obviously avid birders) wandered to the edge of the clearing to try and determine why a Tennessee warbler was up and singing at 10:30 in the evening. They located our bait and soon all campers were gathered to examine the properly-aged roadkill.

Observed:

Ten instances of *audible gagging*.

Seven cases of *reflux*.

Eighteen *'tossed cookies'*. (Samples were carefully gathered and were submitted for analysis.)

Conclusion(s):

These data closely conform to previous observations. Nothing new was learned. Humans possess extremely weak stomachs.

We spent the remainder of the evening carefully observing the camp in order to prepare for tomorrow's first exercise. Finally, our well-honed powers of observation and hearing paid off as Ellie and Nigel discovered a group of six males who were planning a morning hike to the top of the mountain via a remote trail. We quickly crafted our detailed plan of observation.

The First Encounter:

Our plan (like *all* of our plans) was simple, yet masterful. Cassie and Nigel would remain to observe the campsite while Ellie, Herschel, Morris and I would accompany the hikers. When the humans were an eighth of a mile from the camp, Ellie and Herschel began to 'pace' them. (This is a maneuver in which we accompany the hikers – in this case, Ellie on the left, Herschel on the right – and remain unobserved but make no effort to conceal the sound of our footfalls. When the hikers stop, we also stop, after being careful to take one additional step to ensure that the humans are aware of our

Harold

Elvira

presence.) After three quarters of a mile of 'pacing' Herschel and Ellie reverted to the concealed/silent mode to allow the hikers to relax. Meanwhile, at a point on the trail a mile and a quarter from the campsite, I took position behind a large poplar tree. When the hikers approached to within ten paces of my position, Ellie signaled by bird call (blue jay). I then went into a crouch, stepped into the trail, turned to face the oncoming humans, rose to my full height of nine feet and eleven inches, threw my arms into the air, did a *nifty* little dance step and said, **"WHASSUP, DUDES?!"** (Now, I realize that, due to the difference in out languages, this *may* have *sounded* like a growl. In fact, due to the enthusiasm with which I spoke, it *could* have been mistaken for a *ROAR*! But I can't be blamed for that.)

Observed:

For an instant, time seemed to freeze as the hikers absorbed what had transpired before them. Then action began apace. First, there were six, near-simultaneous human scat samples provided. (Unfortunately, since all humans were wearing jeans, none of these were collectable.) Then, in a shower of discarded equipment, the hikers whirled and started back towards the campsite as fast as their pathetic little underdeveloped legs would take them. Herschel gave chase but had to pull up after three strides to keep from getting ahead of them.

Herschel

Meanwhile, Morris, Ellie and I took inventory of the discarded equipment:

Six water bottles

Two camcorders

Three still cameras (one film, two digital)

Five pair of binoculars

Six backpacks

Five hiking boots

Three GPS receivers

Since very little of this equipment was of any use to us, we simply noted the location of each item, the brand names and condition and left them where they lay. There were, however three exceptions:

1. Morris ate one of the hiking boots. Although it had a pleasant aroma of properly-aged roadkill, he found it to be somewhat tough and not entirely to his liking.

2. Ellie thought that the knapsacks could be modified for useful service as handbags or fanny-packs, so she retained two of those.

3. We ALL enjoyed the glazed donuts and chocolate bars. (At least those we could keep from Morris.)

After our former hikers reached the campsite, they breathlessly told their story. Their fellow campers quickly convinced them that their eyes were playing tricks and all that they had seen was an opossum. (After hearing this, all the researchers began to call me 'Haropossum'. I became so angry, I didn't speak to anyone for twenty minutes!)

Conclusion(s):

These data closely conform to previous observations. Nothing new was learned. Human are the SLOWEST vertebrates in the forest.

The next encounter was, as happens from time-to-time, an unplanned and spontaneous experience. Yet it was an encounter which provided more information than we have been able to gather in quite a while.

The Second encounter:

At dusk on Saturday, a female (we later learned her name was Alicia) left her campsite to answer a 'call of nature'. Cass was in her sector at the time and accompanied her (unobserved/silent mode) on her trek. When Alicia found just the right spot and assumed the position to 'take care of business', Cass stepped into the open in front of her, gave her a friendly smile and uttered a low, "Whoop!" (our language for, **"Gotcha!"**)

Observed:

ALL investigators were extremely impressed by Alicia's speed and agility. According to Herschel (Expedition Statistician), she covered the one hundred and nine yards back to her campsite in ten seconds flat! (This is even MORE impressive considering the fact that Alicia accomplished this with her jeans *around her ankles*!) At one point, we looked on in horror as she was headed directly for a large blackberry patch. If she encountered those vicious briars at the speed she was moving, she would seriously injure herself! As it happened, we needn't have worried. NONE of us had EVER seen a human – or *any* animal for that matter - run across the TOPS of blackberry bushes!

Upon reaching her campsite, Alicia breathlessly related her story. Her fellow campers quickly convinced her that her eyes were playing tricks and all she had seen was a raccoon. (All researchers began referring to Cass as 'Cassaraccoon', whereupon she became so angry she wouldn't speak to anyone for twenty minutes!)

An excellent scat sample was carefully collected and submitted for analysis.

Conclusion(s)/Recommendation:

After close examination of all data, we learned that Alicia's demonstrated speed was only four percent slower than the legendary Simon ('the Slug') Snailfoot, the slowest sasquatch known to history. Congratulations, Alicia girl! You are now, officially, the fastest human we have ever clocked and a part of sasquatch history!

It is hereby strongly recommended that we closely study Alicia's technique for running across the tops of briar patches. If we can master this maneuver, it could well be used to our advantage in the future.

After this encounter, we retired to the deep forest to allow the humans to enjoy their supper while we planned our next exercise, 'The Serenade'.

The Serenade:

This event, also known irreverently as 'The Whistle in the Thistle' is employed as an after-dinner entertainment to reward our human subjects as much as anything else. Herschel and Ellie would observe the campsite while Nigel and Cassie would perform a serenade of whistles, whoops and howls. Morris provided percussion accompaniment with tree and rock knocking. Meanwhile I ran noisily through the nearby forest breaking large limbs and small trees. (This is NOT my favorite part of any expedition as I invariably get pine sap all over me and my hands are sticky and yukky for a week!)

Observed:

Shock   100 percent

Awe   100 percent

Conclusion(s):

These data are consistent with previous results. Nothing new learned.

After our performance, we again retired deeper into the forest to compare notes and discuss our next move. It was unanimously decided that nothing further of any significance would be learned from this group and it was decided to proceed with 'Operation Termination'.

Operation Termination:

After our concert, we allowed our humans to relax and retire to their tents after a full day of activities. Then, at 1:30 a.m., all six of us entered the camp, making no effort to conceal our footsteps, and walked among and around the tents, occasionally grunting, snorting and picking up and noisily discarding various items of camping equipment. (It never ceases to amaze us that, during this exercise, no human is anxious to leave his tent and join us.) After twenty minutes, we quietly retired to the forest and took our positions to observe the humans' camp-breaking techniques.

Observed:

The last club member entered his vehicle in forty-five seconds. The last vehicle squealed onto the paved road in four minutes. (Not a bad time for transiting three miles of rutted logging track!) We left the tents and other equipment where they were. Morris took care of all remaining comestibles.

Conclusion(s):

Although their departure was somewhat more expeditious than the average, our observation did not significantly differ from previous expeditions.

Overall:

Some things, thanks mainly to Alicia, were learned from this expedition. However, we have studied humans for so long that there is, frankly, little left to learn about them.

We hope to have the laboratory results back within two weeks and have the final report out within a month.

Pictures and maps will be made available as Nigel finishes drawing them.

NO HUMAN WAS INJURED DURING THE COURSE OF THIS EXPEDITION. (However, there was minor to moderate suspension damage to nine vehicles during the exit phase and six mufflers were lost.) (We are still not certain as to what use the humans make of 'mufflers' but we find that, when they are banged together or hit with sticks, they make sounds that we find pleasing.)

/s/ Harold Harefoot

Chief Investigator and Lead Scientist

We Are But Parts of the Whole

Ok, we have enjoyed our trip into fantasy, now is time to return to reality.

In our endeavors on behalf of the sasquatch people, we often forget there are others on their own journey into this world of the soon to be known. We are all in our own position in this vast world. We all have our own work to do and our own goals to accomplish. It is most important to remember that we are all related. We all want the same end. Some of us say lo here... others say lo there, but the point is, we are all investigating the same enigma. Never, please forget that.

Chapter 17

The Many Parts of the Whole

By Thom Cantrall

It was late on a pleasant if damp early spring day when the lady and her companions arrived at the remote lake so near the cold and broad Pacific Ocean. Their quest was for the great and elusive wild people that inhabit this far western land. They knew not what they would find, if anything, in this isolated spot but the time had come to take the next step on their personal journeys of discovery.

They were here today because I had sent them here. Only days prior, she had asked me where she might go in her area to have an opportunity to encounter or, at least, interact with the enigmatic sasquatch. I inquired, at her behest, of my teacher and he told me to send her to this remote and lonely spot.

"Mom," the young man asked, "since we've got our camp arranged and there's still light, why don't we hike around the lake before we fix dinner? Maybe we will get to see something important."

The hike was eventful. Evidence of the presence of their hosts was found in the form of a very pleasant stick formation or glyph and a signature footprint in the soft trail. The truth of the host's proximity was found on their return to their camp after about twenty minutes on their evening hike. Carefully placed atop the small, portable gas grill were two tiny plastic caps from the gas canisters used prevalently as a fuel source for stoves and lanterns. Somehow, in the relatively few minutes of their short hike one of their hosts had entered their camp and left a calling card...

In the same time frame Jeff, from the Portland/Vancouver metropolitan area contacted me with a similar request... "Where can I go to have an interaction?" I asked and was told to send him a very specific area on the South Fork of the Lewis River. He heeded this suggestion and had a very intimate conversation with the local denizens... "You, know," he stated to me, telling me something I had not, until that moment known, "I knew you were not just looking at the map and picking a spot because you sent me to the South Fork of the Lewis River. That is a local name only. All the maps and literature call it the East Fort of the Lewis." Well, I knew I'd not looked at a map to send him here and I'd never been there myself, but I had a very clear vision of it in my mind, even to the point of being able to give him landmarks to look for in the area he was to investigate...

There is a spot in southwestern British Columbia, Canada where a very active and important clan resides. Brian began his investigations there early in 2012. He is very fortunate to have an "on the ground" mentor who has years of experience in this area with the local residents of the woods and mountains. What he has been shown in this short span of time is nothing short of miraculous! The scope and range of stick art glyphs left for him in this area are simply spectacular. Daily when he ventures forth, there is, invariably, a new discovery awaiting him.

He has listened to their chatter as these beings talked among themselves while he and is partners stood so close yet so very far away. Were they possibly discussing advisability of showing themselves to these curious hairless men?

Shadows have passed from tree to tree, teasing and tantalizing the investigators. Pictures are taken of remarkable tree structures and videos are done of their handiwork… and the evidence mounts…

It was a warm summer day in southeast Oklahoma and Arla with her friends were building a Medicine Wheel in the Indian way… The center stone has been carefully placed and the construction begun, only to have the stone repositioned by one of the local entities that live here. Long the ladies labored in the heat and humidity that is this part of the country in the warm months when a break was called. Soon it was noticed that they had company. "We are taking pictures," one lady stated. "If you wish to remain you may, but if you don't want to be in the picture, you'd better move." He didn't move and Blue, an adolescent male that resides there joined the body of evidence.

There was another time, earlier when the Old White One was in the line of fire of the pictures and the same warning was given him. He chose not to stay for the picture and an entirely new and different type of evidence emerged. That the image of this large fellow remains as a shadow is more than remarkable... it is astounding. I have known of the existence of this property in these beings but I had never personally witnessed it but here is photographic evidence of its existence.

It was on the Equinox in 2011 when the rationale for the repositioning of the hub-stones of the Medicine Wheel became apparent as the sun rose and shone through the large X-formation they had made prior to the construction of the Wheel and directly in alignment with the center of the Wheel.

It was later on this same day, September twenty-second, 2011 that Arla was sitting on her chair in the vicinity of this Medicine Wheel and shooting a bit of video using her telephone only so that she could show me something of the nature of her land. Some two minutes into that video there appears, center stage, a large adolescent male accompanied a juvenile and he appeared to be carrying a third, very young individual on his chest. This bit of video is sacred to me as it is to Arla. I have shown it to a very chosen few with her permission. There are those who would argue it and denigrate it and its maker and we simply will not allow that.

This land where Arla resides is magic. It is the land of Indigo and of his older sister, Meredea. It is very near where Rahjahsay the blue-eyed daughter of the clan leader in Brian's area came this fall to meet her mate. The wise and ancient Kashima lives here as well.

It is through Kashima that I met my own Teacher, Akanneesha. He conveyed to Arla a message for me... a message of where I was to look for and meet my Red Striped companion and protector. This land is an abundant land... it teems with evidence. It is evidence there for all who would understand or who would listen.

Far into the southeast there are two special people that are learning their way in this world of strange beings and stranger powers. Near Atlanta, Georgia, Alex arranges his microphones to best advantage to catch the utterings of the denizens of the night that live in his area. His efforts have yielded much fruit of very high quality and we hear the sounds our large friends make. To his south, Keith and Cathy are laboring to understand the denizens of their lands...

In the far north into Ontario, Canada, a new investigator adds his evidence to the body as he walks the lowlands and ridges of that great wild place. His tracks and trackways are shedding light on movements and activities we have heretofore only guessed at in wonder.

Far to the west, on a cold, damp February day in 2012, a young man was told of something remarkable in a drying lakebed near the university town of Eugene, Oregon. He had the presence of mind to call together a team to investigate this London Trackway and the efforts of Toby and others like Thom and Beth and some who do not wish to be mentioned, over a hundred and twenty individual tracks were measured, analyzed cataloged and preserved. This effort will to down as one of the most significant finds to embolden the body of evidence.

One of the primary investigators on this project was called in from his position at Idaho State University and Dr. Meldrum's expertise was instrumental in verifying the legitimacy of this find. He is a world class expert in his field. He is the "go-to" man where questions of the motion and method of these people are concerned.

As is now obvious, many people serve many places in bringing this enigma into the light. Personally, my hope is that sasquatch will never be "proven" to science. We, as a people, have a terrible history with indigenous peoples. We have not learned anything in the thousands of years of our coexistence with others. The chaos that would result from the proof of their existence in our midst would destroy all efforts to understand them. That they may be allowed to live freely and happily is my fondest wish.

Epilogue

A Time To Answer the Call

It's TIME!

The time has come. There are no more "ape groups" nor are there "human groups". If we are to persevere and bring forth a new species into the world (Homo sapiens hirsutii is my choice for a name... you may have your own), we have got to drop such external trappings and embrace what is... not what "may be"... Whatever these beings are, THEY ARE! That they are real is no longer in question. As my friend, Cliff Barackman stated, "They are not fully animal nor are they fully human. They are SASQUATCH!" That is their place in our world.

It matters not to them how we consider them. What is, *IS*... nothing more, nothing less. The arguments over Science vs. Spiritualism must end today. It is not important... they are what they are. **NO ONE**, not me, not the finest scientist in the world has all the answers... nor are they likely to have them in the near future. I have my beliefs. So be it. Others have their own belief system. They are entitled to those beliefs as I am entitled to mine. I can agree or disagree as I will. That is proper. It is not proper for me to denigrate, belittle nor disrespect another for how they believe. The time has come to agree to disagree and move forward. Their nature and capabilities are NOT IMPORTANT today. Their SURVIVAL is. Time will show us their capabilities. Time will show us their nature. Stop bickering now.

There is a significant study pending that will answer many questions and pose many more, as is the wont of such studies… be that as it may. We must learn from this study and move forward. We must take the data from this study and use it to gain protection for their very lives. It is imperative on us to do so and to do so with haste and vigor. There are factors that will be made known that will rock millennia old beliefs to their very core. That, also, is at it is.

There is only one division that will remain within the nonce… should we kill one "for science" or should we not. I will remain adamant on this point. To murder one of these beings would be so wrong on so many levels as to be beyond the realm of belief.

Even if they are found to be merely the basest of animals, they do not deserve death. I am a hunter. I have been my entire life past my fourth birthday. This said, I have never before nor do I now ever kill for sport. I kill an animal for food, to protect myself or others or to maintain a healthy population level… there are no other reasons to do so. All creatures fit into the ecology of the biome for one purpose or another. It is not our place to judge this fact.

Please, my friends, come forward and pledge your support to the emergence of our friends of the forest… our Primal People… It is time to stand up and be counted.

When I first met Akanneesha I asked him what his message was for our species. I mean, I know that there had to be a reason he was going to all this trouble to contact me and to teach me enough to be able to work with him, I knew he had to have a reason for it…

He answered: "The answer to that is simple… RESPECT… Human do not have respect in their lives. They do not respect themselves therefore, they cannot respect others. They do not even respect the place they must live. They are busily destroying their own home and dirtying their own nests."

"Do you realize that for the first time in history, man has the ability to destroy the world without firing a shot from his guns? He is killing the large predators and this will destroy the balance that the creator has established. When the top predators are gone, the lesser species flourish and this will flood the world with creatures that have no control on their population or appetites. This will set into motion a chain of events that will not be possible to stop."

I cannot improve on that.

Thom Cantrall

Made in the USA
Middletown, DE
17 June 2019